T0211882

Health and Girlhood in Britain, 1874–1920

Palgrave Studies in the History of Childhood

Series Editors: **George Rousseau**, University of Oxford, **Laurence Brockliss**, University of Oxford

Palgrave Studies in the History of Childhood is the first of its kind to historicise childhood in the English-speaking world; at present no historical series on children/childhood exists, despite burgeoning areas within Child Studies. The series aims to act both as a forum for publishing works in the history of childhood and as a mechanism for consolidating the identity and attraction of the new discipline

Editorial Board: Jo Boyden, University of Oxford, Matthew Grenby, Newcastle University, Heather Montgomery, Open University, Nicholas Orme, Exeter University, Lyndal Roper, University of Oxford, Sally Shuttleworth, University of Oxford, Lindsay Smith, Sussex University, Nando Sigona, Birmingham University

Titles include:

Hilary Marland
HEALTH AND GIRLHOOD IN BRITAIN, 1874–1920

George Rousseau
CHILDREN AND SEXUALITY
From the Greeks to the Great War

Palgrave Studies in the History of Childhood
Series Standing Order ISBN 978–1–137–30555–8 (Hardback)
(*outside North America only*)

You can receive future titles in this series as they are published by placing a standing order. Please contact your bookseller or, in case of difficulty, write to us at the address below with your name and address, the title of the series and the ISBNs quoted above.

Customer Services Department, Macmillan Distribution Ltd, Houndmills, Basingstoke, Hampshire RG21 6XS, England

Health and Girlhood in Britain, 1874–1920

Hilary Marland
Professor of History, University of Warwick, UK

First published 2013 by
PALGRAVE MACMILLAN

Palgrave Macmillan in the UK is an imprint of Macmillan Publishers Limited, registered in England, company number 785998, of Houndmills, Basingstoke, Hampshire RG21 6XS.

Palgrave Macmillan in the US is a division of St Martin's Press LLC, 175 Fifth Avenue, New York, NY 10010.

Palgrave Macmillan is the global academic imprint of the above companies and has companies and representatives throughout the world.

Palgrave® and Macmillan® are registered trademarks in the United States, the United Kingdom, Europe and other countries.

ISBN 978-1-349-46037-3   ISBN 978-1-137-32814-4 (eBook)
DOI 10.1057/9781137328144

This book is printed on paper suitable for recycling and made from fully managed and sustained forest sources. Logging, pulping and manufacturing processes are expected to conform to the environmental regulations of the country of origin.

A catalogue record for this book is available from the British Library.

A catalog record for this book is available from the Library of Congress.

To my grandmothers, Mary Smethurst and Alice Marland, and my mother, Joan Marland

Contents

List of Figures   viii

Acknowledgements   x

Introduction   1

1 Unstable Adolescence: Medicine and the 'Perils of Puberty'
   in Late Victorian and Edwardian Britain   15

2 Reinventing the Victorian Girl: Health Advice for Girls in
   the Late Nineteenth and Early Twentieth Centuries   42

3 Health, Exercise and the Emergence of the Modern Girl   86

4 Girls, Education and the School as a Site of Health   122

5 The Health of the Factory Girl   155

6 Conclusion: Future Mothers of the Empire
   or a 'Double Gain'?   189

Notes   197

Bibliography   247

Index   265

Figures

1.1 Two 'before and after' portraits of a young girl, Miss C., with the condition anorexia nervosa. From William Gull, 'Anorexia Nervosa (Apepsia Hysterica, Anorexia Hysteria)', *Transactions of the Clinical Society of London*, 7 (1874), 22–8, Case C, facing p. 26 (Wellcome Library, London)   32

2.1 Carter's Little Liver Pills for headaches, biliousness, torpid liver, constipation and indigestion: Display card, c.1910 (Wellcome Library, London)   50

2.2 'Physique and character'. From Mary Humphreys, *Personal Hygiene for Girls* (London, New York, Toronto and Melbourne: Cassell, 1913), p. 15 (author's collection)   63

2.3 Gymnastic dress illustrated in 1893. 'Perambulating Wind Mill'. From Ellen Le Garde, 'Outdoor Sports for Girls', *Our Own Gazette*, X: 117 (September 1893), p. 106 (Modern Records Centre, University of Warwick, MSS.243/5/4: YWCA/Platform 51)   70

2.4 Venus de Medici compared with a body produced by tight-lacing. From Malcolm Morris (ed.), *The Book of Health* (London, Paris and New York: Cassell, 1883), p. 498 (author's collection)   74

2.5 'The result of too hasty toilet. Notice the safety pin'. From Mary Humphreys, *Personal Hygiene for Girls* (London, New York, Toronto and Melbourne: Cassell, 1913), p. 95 (author's collection)   76

3.1 Lawrence Liston, 'Bicycling to Health and Fortune', Part II, 'The Rider', *Girl's Own Paper*, XIX: 939 (25 December 1897), facing p. 198 (The British Library)   99

3.2 E.M. Symonds, 'Physical Culture for Girls', *The Girl's Realm Annual* (November 1898–October 1899) (London: Hutchinson, 1899), p. 62 (The British Library)   101

4.1 Leg and balance movement (note the poor environment in which exercise took place, and the oversized dress worn by the girl in the photograph). Mrs Ely-Dallas, Organising Teacher of Physical Exercises under the School

Board for London, 'Physical Education in the Girls' and
Infants' Departments in the Schools of the London
School Board', in *Special Reports on Educational Subjects*,
vol. 2 (London: Her Majesty's Stationary Office, 1898),
Figure 7 (author's collection)   126

4.2   The North London Collegiate School Gymnasium. From
E.A.L.K., 'The North London Collegiate School for Girls',
Girl's Own Paper, III: 122 (29 April 1882), facing p. 494
(The British Library)   132

5.1   New YWCA Institute Buildings, Liverpool. From *Our Own
Gazette*, I: 11 (November 1884), pp. 126–7 (Modern
Records Centre, University of Warwick, MSS.243/5/1:
YWCA/Platform 51)   163

5.2   Drill competition at the Guildhall, London, 1924. From
YWCA. A Review 1924, p. 2 (Modern Records Centre,
University of Warwick, MSS.243/2/1/9: YWCA/
Platform 51)   172

5.3   At work on munitions boxes. From *YWCA. A Review
1916*, p. 16 (Modern Records Centre, University of
Warwick, MSS.243/2/1/2: YWCA/Platform 51)   179

5.4   YWCA munition workers' canteen. From *YWCA. A Review
1914–1920*, p. 5 (Modern Records Centre, University of
Warwick, MSS.243/2/1/5: YWCA/Platform 51)   181

6.1   Mary Humphreys, *Personal Hygiene for Girls* (London,
New York, Toronto and Melbourne: Cassell, 1913),
frontispiece (author's collection)   194

Acknowledgements

This book explores the ways in which debates and advice on girls' health which emerged in the last quarter of the nineteenth century contributed towards the creation of a new cultural category of girlhood. Female adolescence was depicted as full of potential for young women, as they garnered and honed their health and vitality, which enabled them to enter into new activities in the workplace, school, sport and recreation, but also potentially put at risk their vulnerable bodies and minds as well as their role as future mothers. It is interesting – and also troubling – to reflect upon how relevant such issues remain in considering anxieties about the health of young women in the early twenty-first century. Even as girls outrank boys in their examination results and break through the glass ceiling in many careers and professions, concerns are still expressed about their disproportionate susceptibility to anxiety, depression and eating disorders, highlighting again and again the apparent vulnerability of the adolescent female body.

In carrying out research for this book, it has been a great pleasure to draw extensively on my 'local' archive, the Modern Records Centre, based at the University of Warwick, a wonderful working environment with knowledgeable and helpful staff. The interlibrary loan service at Warwick University Library responded to my numerous requests for material with great efficiency. I am grateful too for the excellent resources offered by the British Library, the Wellcome Library, the Women's Library and the Guiding UK archives, and would especially like to mention the staff at the British Library Newspapers at Colindale for their care in producing and reproducing reams of material. I had a warm reception on my visits to the library of the North London Collegiate School for Girls. Thanks are also due to the very helpful staff in the images and reproductions departments at the Wellcome Library and the British Library for granting permission to reproduce illustrative material, and to the Modern Records Centre and the Young Women's Christian Association (YWCA)/Platform 51 for allowing me to use images from the YWCA archive. During the years this project has been in development, I have had very valuable feedback from audiences at numerous workshops, seminars and conferences, notably at McMaster University in its early stages, in Exeter and Manchester, and from my own colleagues

Acknowledgements xi

at Warwick. Rima Apple advised and suggested ways of structuring my book in its nascent days and provided comments in the closing stages. Catherine Cox read sections of the book and made valuable suggestions and has been a generous friend and collaborator. Vicky Long shaped my thoughts on factory work and many other issues. Colleagues and the postdoctoral fellows and postgraduate students based in the Centre for the History of Medicine and the History Department at Warwick have created a great community, in particular Roberta Bivins and the excellent Tracy Horton, always at the end of the email, and also David Arnold, Angela Davis, David Hardiman, Sarah Hodges, Maria Luddy, Molly Rogers, Claudia Stein, Mathew Thomson and Sarah York and my ex-colleagues and friends Margot Finn and Colin Jones. Research assistance was provided at various stages of the book by Jane Adams, Atsuko Naono and Daniel van Strien. Clare Mence at Palgrave Macmillan has been a source of sage advice and calm in getting the book into production, and Flora Kenson a meticulous project manager. The series editors Laurence Brockliss and George Rousseau have been enthusiastic throughout, and I am delighted to be part of their new series. Last, but far from least, my family have been a source of fantastic support, and have never suggested for a moment that I was taking far too long to complete this book, so many thanks to Sebastian, Daniel and Sam. My mother, sadly, did not see its completion, but I would like to dedicate this book to her and to two Lancashire 'factory girls': my grandmothers.

Introduction

In 1891 Dr Gordon Stables prefaced his volume *The Girl's Own Book of Health and Beauty* with the chief aim and objective of his work:

> I desire to teach my girl readers how to be healthy, because there can be no beauty without health. Brightness of eye, clearness of complexion, and happiness of expression, belong only to the possessor of health ... my main object all through has been to persuade the reader that health – bounding saucy health – is the fountain from which all true beauty springs.[1]

Providing more insight into how girls might achieve this state of true health and beauty, Stables related the tale of Mary and Elsie, 'innocent, simple, servant lassies'. While Mary was healthy, happy and rosie, 'an interesting and rather pretty girl', Elsie in contrast was 'nervous-looking, rather languid and pale ... an anaemic subject'. 'I always feel so tired', Elsie proclaimed; a cry, according to Stables, 'of the rich as well as the poor'.[2] Obedience to the laws of health as spelt out in his book, would, he declared, restore health and vitality. These laws comprised attention to diet, which was to be temperate and taken at regular hours; warm salt water bathing; plenty of exercise, ideally with friends; recreation, social intercourse, reading and amusement; and, most important, a sense of purposefulness. 'In your calmest, quietest hour of the day – and that will probably be first thing in the morning – give yourself up to thinking and planning something that shall benefit and restore you.'[3]

By the time Gordon Stables, ex-navel surgeon and author of adventure stories and reams of popular health advice literature, produced his book, he was contributing to a growing trend in writing for audiences comprised primarily of girls. Books and magazines for girls were becoming

1

standard and profitable fare for publishers, and a good deal of this literature focused on health and how to achieve it. Stables' book also fed into
a tendency for doctors to engage increasingly with the broader public,
propagating their medical theories and ideas on treatment and health
improvement across a variety of print media, advice literature and pamphlets, popular periodicals, magazines and health journals. Much of
this information was directed towards young women. The publication
of *The Girl's Own Book of Health and Beauty* came ten years into Stables' 30-year stint as health columnist to the *Girl's Own Paper* (*GOP*) and
apparently inspired his other books on health.[4] In his book and advice
column Stables described young women as able and energetic, and as
having the potential to further augment these attributes through healthful activities. Free from jargon and speaking in plain language – though
laced with cloying sentimentalism and patronising vignettes – Stables
engaged vigorously with the themes that frame this book. He presented
health as a desirable and achievable object. He saw girlhood as a distinct
and important phase of life. And he recognised opportunities in linking
the two: girls ideally would not only be the possessors of true beauty,
they would also be in blooming, vital health.

Social and cultural historians have presented the opening decades of
the twentieth century as being marked by a major shift in ideas about
what constituted a healthy female body, a process dovetailing with the
increased visibility of girls, in part due to their employment in wartime
industries and paid labour more generally and their interest in physical
culture, sport and new leisure activities. The visibility of the 'flapper'
in particular prompted concerns about declining standards of girlhood,
as the modern young woman opted for boyish hairstyles and clothing,
and became a victim of the fashions for a slender physique, dancing
and cigarette smoking.[5] However, there are clear indications that new
models of healthy girlhood were already being established in the late
Victorian period, symbolised by Stables' bounding, healthy girls. During
this era, young women started to move out of environments dominated by family obligations and domesticity into new roles and more
public spaces – the workplace, school and social, cultural, political and
sporting arenas – becoming the subjects too of criticism about declining standards of behaviour and rejection of their natural femininity,
which accompanied these more public and demanding roles. By the
turn of the century the sporting woman, for instance, had become 'a
compelling image ... [and] exemplified both female physical emancipation as well as ambivalence about women's entry into an essentially
masculine world'.[6]

While the purported limits of girls' biological and psychological make-up were construed by doctors and many other observers as representing barriers to change, other commentators highlighted the ways in which enhanced health enabled girls to take advantage of new opportunities and presented new visions of the health that girls could realise by adhering to a few straightforward rules and practices. The characteristics of the 'modern girl' with her 'modern body' were, if not realised, then at least aspired to by the 1880s and 1890s – a healthy body, honed by sports and physical exercise, clothed in modern attire appropriate for movement and flexibility, with an exuberant personality to match. This book explores how within a few decades the wilting, pallid, awkward mid-Victorian girl, who featured heavily in novels of the period as well as in medical discourse, was visualised increasingly as fit, radiant, filled with energy, cheery and outgoing, as she bounded into new and challenging social and occupational roles. It will also examine the complexities, limits and mixed responses to this transformation. Improved health and vitality was equated by some doctors and social commentators with the loss of the qualities of femininity, compassion and modesty, resulting in a new breed of girls, loud and graceless, wielding hockey sticks, speeding on bicycles and wearing unbecoming masculine attire.

The notion of girlhood as a separate stage of existence with its own values, interests and readership potential evolved from the 1870s onwards, as publishers began to identify readers who were neither children nor adult women. Sally Mitchell has traced how in the late nineteenth century the emphasis on home life diminished, as increasingly girls occupied the semi-public spheres of work, education, sport and leisure, and featured increasingly in novels and self-help literature.[7] The term 'New Girl' was employed by the journalist and writer Mary Anne Broome who, comparing her own pre-1850 childhood, was convinced of 'the superiority of the modern girl' who was more self-assured and recognised that marriage was not inevitably her destiny.[8] Though it was to be the First World War which in many ways decisively 'snapped strings' that bound women and girls to the Victorian age, by the turn of the century large numbers of young women were employed in mills, factories or shops, or worked in domestic service, while the 'white blouse' revolution created opportunities in the retail sector, civil service and local government.[9] In 1901 over half of all women workers were under 25 years of age, and 61 per cent of young women aged between 15 and 24 participated in the workforce.[10] The number of girls attending secondary school, college and university also increased rapidly. After primary education became compulsory in 1870 elementary day schools for girls aged up to 14 proliferated, while middle-class girls were increasingly sent to school rather than being tutored at home, a schooling which by the end of the century was more likely to lead to university study. They also joined clubs and participated extensively in sports and leisure activities as well as charitable work.

Yet, despite this visibility, the notion of girlhood remained ambiguous in many ways. Sally Mitchell has described how publishers identified a new readership amongst girls, but has also suggested that, in terms of popular literature at least, 'girlhood' is difficult to define, and referred at least in part to a state of mind or an aspiration, rather than a legal or age-related concept.[11] The popular weekly *GOP* reflected this ambiguity. Targeted primarily at young adolescents, in terms of its content and choice of illustrations it also projected towards younger and more mature readerships, suggesting that 'girlhood' was a flexible and desirable concept, given the new opportunities that this epoch appeared to offer. For the purposes of this study, girlhood has been taken as roughly embracing the 'teenage' years, albeit at a time when this term was rarely employed.[12] When Gordon Stables referred, for example, to the 'critical' last five years of a girl's teens, he was alluding particularly to the dangers associated with physical change, though he was aware and not unsympathetic to the formation of a culture of girlhood which ideally should be founded on health and healthful behaviour.[13]

The idea of girlhood and its potential varied greatly according to social class, though authors such as Stables – and his fellow contributors to the *Girl's Own Paper* – were keen to speak to rather romanticised versions of working and servant girls as well as their predominantly middle-class readership. For working-class girls around the turn of the century, girlhood came to be strongly identified with elementary schooling and work outside of the home, though many, as recounted by Robert Roberts, were still kept on a tight rein, while 'devoted' daughters became unpaid domestic drudges.[14] Elizabeth Roberts has suggested that the period between leaving school and marriage, the ages of about 14 to 25, bridging 'childhood and independent adulthood', were distinctive for working-class girls and marked increasingly by wage earning and access to leisure activities.[15] The transformation of young women's lives through leisure after 1920, as described by Claire Langhamer, was shared by many girls before this date, as they became involved in activities that enhanced their independence and freedom, whether this was attending a girls' club, walking or cycling in the countryside, or going shopping or to a concert or theatre.[16] The experience of schooling outside the home, meanwhile, marked the commencement of girlhood for many middle-class girls who were to encounter a girl's culture distinct from the home environments that had hitherto so dominated their lives. While the opportunities open to girls able to stay on at school beyond the age of 14 contrasted sharply with those forced to work for a living, girls of all classes participated increasingly in social and leisure activities, hobbies and games. For working-class girls, who were short of time given the pressures of work and home duties, and denied the luxury of costly sports such as tennis, golf and horse riding, this was likely to be the more limited range of gymnastics and games organised by clubs or their place of work. Yet for all girls, while their lives were still bounded by home life and parental authority, to some degree openings were created to move into public or semi-public arenas, 'spheres where her mother had no advice to give', distinctive 'girl cultures' where girls primarily interacted with each other.[17] A space opened up between childhood and marriage and motherhood, and increasingly opportunities arose to fill this with more than accomplishments such as needlework and piano or – as was reputedly the case for working girls enjoying their earnings – hanging around street corners, bolting down buns and stewed tea, and enjoying the dubious pleasures of the music hall or other cheap entertainments.

These shifts in activity and opportunity were paralleled by another set of changes in terms of bodies and physicality, as during the late nineteenth century the age of menarche started to decline. Most medical writers and authors of advice literature agreed that English girls were likely to start to menstruate around the ages of 13 to 16, and it was affirmed that accelerated development of the nervous system caused earlier menstruation amongst wealthy girls, as did the impact of living in towns, poor diet or lack of sleep.[18] Puberty, 'the time when a girl passes from childhood into womanhood', was deemed 'most important in a woman's life', necessitating care and watchfulness, 'a most trying time'.[19] Physical and mental health at puberty, an area of frantic concern on the part of many doctors, was debated with increasing alacrity as girls developed new aspirations, particularly to enter secondary and higher education, which many commentators argued put at risk their prospects for becoming good wives and mothers. As fault lines opened up in terms of the opportunities created for girls of different classes, this was reflected in medical discourse. Worries about the strain of 'brain-work' for middle-class girls contrasted with increased watchfulness – although not necessarily remedial action – regarding the potential physical damage and lifestyle deprivation faced by many working-class girls. However, for girls of all classes, the physical changes of puberty became coupled with broader anxieties as by the turn of the twentieth century 'youth' or 'adolescence' was increasingly depicted as a discrete and significant era for individual, medical and social development, embodying positive signs of growth and development on the one hand and sinister and risky biological and behavioural change on the other. This was associated most closely with the work of US psychologist and educationalist G. Stanley Hall, who 'yoked the fashionable psychology of normative development to distinguish a separate age between the onset of puberty and mature adulthood'.[20] Many medical writers shared concerns about this distinct phase. Eugenicist physician Dr Mary Scharlieb asserted that the most important years in terms of development were between the ages of around 12 and 21, when 'heredity' bears its crop, while 'environment' still had full play, 'and the outline of these two forces is seen in the manner of individual that results'.[21] Others, notably the women doctors affiliated to the Medical Women's Federation, propagated a more positive outlook and minimised the impact of adolescence, suggesting that menstruation, and puberty more generally, was no barrier to the enjoyment of sport. Nor was it an impediment to paid employment or school work.

In recent decades social, cultural, feminist and medical historians have debated and published extensively on women's reproductive health and medical conceptions of the female body.[22] Yet, despite growing recognition by the end of the nineteenth century of girlhood as a discrete social and medical category, little historical work has focused specifically on young women.[23] Obvious exceptions to this have been explorations of anorexia nervosa, chlorosis and hysteria, extreme medical events that make up part of the story of girls' health and challenges to it during this era, but far from the whole story.[24] By the late nineteenth century medicine operated along 'Darwinian-science lines', according to Patricia Vertinsky: 'Although women were held to be victims of their entire reproductive experience, the onset of menstruation and its recurring cycle were believed to be a cause of particular handicap.'[25] Historians such as Lorna Duffin and Maria Frawley, meanwhile, have suggested that medical arguments demonstrating a limited capacity for healthful activity and limited social roles for women contributed to the culture of the invalid amongst nineteenth-century women.[26] Girls were not only liable to incapacity, but concerns about the onset of menstruation meant that 'they must be treated as invalids'.[27]

This book not only shifts the focus onto girls but questions how far they were conceived of as being solely or even largely victims of biological vulnerability. Close exploration of medical and lay publications on the theme of girls' health reveals the extent to which dominant medical discourses about the risks of puberty were debated and upset from the late Victorian period onwards, whether this was by doctors promoting cycling or other forms of physical activity, headmistresses encouraging girls to take on the challenges of the classroom, examination system and the sports field, or welfare supervisors aiming to better the health and lifestyles of factory girls. An exploration of health advice literature, meanwhile, suggests that girls were much more likely to be described as suffering from nervousness or listlessness rather than hysteria, or as being too fat or too thin rather than anorexic. A diverse range of health issues, relating to the everyday practices of hygiene, the management of menstruation, diet, exercise and mental wellbeing, were debated in medical literature and tackled in health advice forums catering for young female audiences, issues that were anchored to concerns about girls' increased involvement in intellectual life and education, employment, and social, cultural and sporting activities. Many of those promoting the pursuit of health in young women suggested that this would open the door to new opportunities to be pursued with resolve and vigour, though this could be tempered by a concern to protect vulnerable female bodies and minds and preserve them for the primary function of future motherhood. The turn of the twentieth century was also critical as an era when interest in girls and youth as a whole was evolving rapidly. Girls were described as becoming modern, and in some ways a force of modernity, with their increased engagement with knowledge and the workforce and their rapidly changing appearance. However, spurred by anxieties about citizenship, Empire and eugenic thinking, anxieties about girls' future roles as mothers also revived, particularly in the build up to the First World War.[28]

Studies of women's health in the late nineteenth and early twentieth centuries have focused predominantly on reproduction and maternity and, to a lesser extent, the dangers women faced as they entered the workplace, particularly the munitions factories of the First World War.[29] Barbara Harrison has gone so far as to argue that between 1850 and 1945 'interest in the health of women centred around a single issue manifested in two principal strands of debate'. The single issue was reproduction, the two strands of debate were the 'health of the physical body of the nation' and the 'social body, and the extent to which women's health could be used as a political strategy to maintain a sexual division of labour in the public domains of work and education and the private domains of the family'.[30] John Pickstone's landmark article on the political economy of medicine in the twentieth century, meanwhile, engaged largely with the experiences of working men, and featured women's place in 'productionist' medicine chiefly in terms of pronatalist campaigns and their task in 'bringing up the reinforcements', as mothers of future generations of workers and soldiers.[31] While reproduction and maternity were undoubtedly issues of vital significance during this period, emphasis on the preservation of women's reproductive health and a historiography which argues that women's health became an issue of state concern principally through the development of maternal and child welfare policies can be questioned.[32] Women were also influenced by changing health ideals and steadily improving services directed toward nutrition, housing and working conditions, and their lives were permeated by advice on birth control, hygiene and fitness. Before the First World War, health, as promoted by Arthur Newsholme, England's Chief Medical Officer to the Local Government Board, was declared a priority of a modern lifestyle, even if this largely focused on educating 'the masses' to improve their health status to combat disease.[33] Women's organisations, ranging from mother's meetings to the Women's Co-operative Guild, meanwhile, took matters into their own hands, creating services and campaigning for better state provision for themselves and their children.[34] As 'the Victorian belief that procreation was the "primary purpose" of marriage gave way to a new emphasis on companionship and sexual fulfilment', manuals advising on sexuality and information on birth control slowly became available to married women of all social classes.[35] Oral history has also shown the extent to which working-class women engaged within their own families, neighbourhoods and communities with the issues of family limitation, sexuality, motherhood, childbirth and child care, passing on health advice and care through informal channels.[36]

Girls have loomed small in most of these accounts, even in terms of their potential capacity as future mothers, though Carol Dyhouse and other historians of education and the social construction of girlhood have focused on the domestic training of young women, and efforts to raise standards of housewifery and childrearing amongst future mothers, particularly as an aspect of the school curriculum.[37] Dyhouse has also engaged vigorously with the 'socialisation' of young women in the late Victorian and Edwardian periods, the shaping of their education and its relationship with feminism.[38] In recent years, an expanding literature has focused on girls' clubs as a means of reforming, restraining and educating unruly and unmanageable urban girls, but the promotion of health by club workers has received little attention.[39] Ellen Ross and others, however, have invoked the hardships faced by working-class girls in Edwardian London, as they were expected to assist their mothers in endless domestic tasks; 'little mothers', 'practical-minded, careworn, vigilant girls' were a common sight in working-class areas.[40] Even as schooling extended for girls, domestic need was recognised and older girls often became 'half-timers', combining the 'double burden' of school and paid work or school and household chores, including large amounts of child care. By the time they married and had their own children these 'little nurses', who took care of the baby of the household and attended to their other siblings, had become 'weary of motherhood'.[41] Even middle-class girls were likely to take on more duties around the home in response to the shortage of domestic servants during and after the First World War. Nutrition tended to be substandard amongst working girls, though middle-class girls also often suffered ill health and poor physical development.[42] Living conditions remained grim for many working-class girls and overcrowded housing was commonplace well into the 1930s, and as late as 1934 the Chief Medical Officer alluded to the high incidence of sickness amongst girls and their liability to disorders such as anaemia and tuberculosis.[43]

This volume develops the theme of girl's health in a variety of settings – school, the workplace and girls' clubs – and also considers the ways in which medical discourse and advice literature shaped and informed discussions on their health and wellbeing. In addition to medical texts and journals, household medical guides and prescriptive literature aimed at girls, health periodicals and magazines, this study will focus on the activities and archives of agencies that included amongst their objectives the promotion of girls' health and welfare, notably the Young Women's Christian Association, guiding and sporting organisations, factory welfare supervisors, and associations of headmistresses and teachers.[44] Such organisations generated a good deal of literature on girls' health, exercise, hygiene and nutrition in the form of pamphlets, periodicals and specialised handbooks, reports of meetings and conferences, minutes, annual reports and mission statements. Linking archive to advice literature provides insight into how prescriptive literature drew on practical experience, or, alternatively, in a small number of cases at least, transformed into practice. Many individuals combined writing on girls' health with work as headmistresses, school medical superintendents, gymnastics teachers and welfare supervisors, resulting in a proliferation of 'expertise' on this topic, and transferred their impressions and their acquired knowledge into their published work. Letter pages in girls' magazines and occasional pieces written by girls themselves, meanwhile, highlighted areas of reader concern, as well as the ways in which girls themselves engaged with health matters. Exploration of a wide cross-section of organisations and agencies – many with overlapping constituencies of experts – produces a more complete picture of the practices of health in a variety of contexts, allowing us to assess the transmission and, albeit more rarely, the impact of medical debates, ideas and advice.

The starting point of this study is a clearly defined one: 1874, a specific date marked by a specific and well-known debate. In April 1874 the distinguished psychiatrist Dr Henry Maudsley published an article in the *Fortnightly Review* which predicted a gloomy future for high school and college girls unable to bear the 'excessive mental drain as well as the natural physical drain' caused by their studies. These would conclude, he asserted, in mental, moral and intellectual deficiencies, and an imperfectly developed reproductive system.[45] His article elicited a swift and forceful response from England's first female physician, Dr Elizabeth Garrett Anderson, who argued that good schools and attention to health-promoting activities encouraged, rather than diminished, vitality in young women.[46] Maudsley's ideas, though far from new, were to be hugely influential and enduring in shaping medical discourse and broader debates on girls' capacity to undertake higher education and their future role as mothers, in underlining the risks of puberty to mental and physical wellbeing, and in fanning debates about girls' liability to hysteria, chlorosis, eating disorders and a pantheon of menstrual disturbances. These debates, however, were to become much more than clashes between 'feminist reformers and Darwinian doctors'.[47] While the views of some doctors can and have been read as wholly prejudicial to women's aspirations, notably, to undertake serious brain work, in other cases they advocated moderation and caution rather than prohibition. For some, the promotion of healthy motherhood did not necessarily run counter to the possibility of intellectual, physical and cultural enhancement, though for others it clearly did.

Medical debate was paralleled by a rather different public controversy, which fuelled notions about girls' aspirations and credentials as future wives and mothers, and questioned the desirability of the new brand of English girlhood. The novels and periodical literature explored by Sally Mitchell have provided an overwhelmingly positive account of the modern girl's prospects, as she bounded forward into a range of interesting and physically challenging spheres.[48] 'The Girton Girl', for example, was described in an 1895 edition of *Atalanta*, as 'a healthy young girl ... She rode, she sang, she swam, she danced, she played tennis ... she could construe one of the hardest bits of Aeschylus, and could prune a fruit-tree with equal grace and readiness. But one thing she never did – she never flirted.'[49] This assessment stood in stark contrast to journalist Eliza Lynn Linton's attack on 'the Girl of the Period', which first appeared in the conservative and widely read *Saturday Review* in 1868. Linton's article, regarded as one of the *Saturday Review*'s most sensational pieces, had an impact well beyond its initial publication, and castigated girls for defying conventions of femininity and flouting their prescribed roles at the cost of the 'tender, loving, retiring or domestic' behaviour proper to English womanhood.[50] 'The Girl of the Period is a creature who dyes her hair and paints her face... a creature whose sole idea of life is fun.'[51] She was inclined to use slang and bold talk and to 'general fastness', was indifferent to duty, useless at home, dissatisfied with the monotony of ordinary life, unlikely to marry and, if she did, would merely spend her husband's money and treat her children like 'a stepmother'.[52] Though Linton prefaced an 1883 reprint of this article with a remark about the ill blood it had caused, she apparently felt no need to temper her views, and her accusations in many ways prefaced attacks on the New Woman, who was described in similar terms during the latter part of the century. Opponents of the intellectually and physically ambitious New Woman criticised her for transgressing boundaries with regard to activities as diverse as higher education, political engagement, cycling and competitive sport, which were described as likely to compromise femininity, motherhood or both.[53] 'Girton Girl' was accompanied into the twentieth century by other emblems of girlhood, several of whom were far from positive – girls with 'bicycle faces', who overdeveloped their muscles and became obsessed with sport, were tense and averse to their domestic roles; rough and vulgar 'hoydens'; 'bachelor girls', who sacrificed femininity to ambition and modernity; and 'factory girls', florid and slapdash, whose poor morals and silliness put their health and character at risk. Several of these girl types will be explored in the following chapters, alongside the impact that their biological make-up and their behaviour was purported to have on their health and wellbeing.

The debate on the psychological and physiological nature of women and its fit with the environment and social order was a long-standing one. Arguments urging equality of educational opportunities dated back to Mary Wollstonecraft's writings in the late eighteenth century, and so too did theories suggesting that women were biologically and psychologically incapable of taking up these opportunities.[54] However, the debate took on new meanings in the late Victorian period, as doctors spoke to emerging concerns about women's role and place in society and domestic life. Medical theory entered the political arena, where it connected powerfully with such themes as women's capacity to engage in politics, to study and to enter public and professional life.[55] For some medical spokesmen, women attempting to move beyond the constraints of domesticity – notably the New Woman or the suffragist – were regarded as strange neurotic anomalies, disrupting the social order and family life and resisting their biological destiny of marriage and motherhood. As obstetrics and gynaecology were forged as discrete and specialised areas of medical practice during the nineteenth century, such women faced a barrage of medical literature undermining their attempts to change their lives. At times, the two specialties opposed each other in claiming expertise over women's health and a share in what was recognised to be a new market opportunity in treating female health problems, but they united in defining links between female nature and the mental and physical risks associated with each stage of the life cycle and in particular the reproductive process.[56] Doctors such as Edward Tilt, Henry Maudsley, John Thorburn, Thomas Clouston and James Crichton-Browne became vocal representatives of a growing band of authorities on the diseases and mental disturbances of women. Until the late Victorian period such writing on female disorders was largely confined to medical journals and textbooks.[57] However, in the last quarter of the century doctors 'no longer content with the specialist pages of medical journals...promoted their views in popular journals that would find their readership in middle-class households', as in the case of Maudsley's influential piece and Garrett Anderson's robust response in the *Fortnightly Review*.[58]

Yet it was not only specialists and elite doctors who expressed views on the health practices and problems of women and girls. Others, in their professional writings and in popular journals and prescriptive literature, grappled in varied ways with the issue of enabling young women to take on new roles without compromising their health and wellbeing. Already by the 1880s and 1890s such media gave voice to a wide range of opinions, including those of feminist doctors, but also medical practitioners such as Gordon Stables, who cornered the market in giving belt and braces advice to young women on how they could improve their health and the quality of their lives. Health periodicals also engaged in energetic debates on the pluses and minuses of female education, the physical training of young women, women and cycling, diet and body weight, hygiene and dress reform. Within a decade of Maudsley's doom-ridden pronouncements on the evils of female education, health guides for the general public were suggesting that 'We need strong, healthy, vigorous women, and not fragile, fainting, insipid creatures.'[59] By the early twentieth century, female eugenicists, a small but arguably influential group, were producing a distinctive and complex analysis of the relationship between girlhood and health, which created a vision of the potential contributions of young, well-educated and energetic women to public life as well as private, family wellbeing. While this group saw the future of the race as being paramount and largely in the hands of young women, mothers to be, they also suggested that girls could and should involve themselves in a range of activities until they took on the duties of home management, marriage and motherhood, conceiving health and vitality as being in a symbiotic relationship.

The following chapters suggest that the late Victorian period through to the 1920s was marked by significant changes in attitudes towards the health of young women as well as in practices of health. The book explores medical literature on puberty and the particular medical problems of girls' adolescence in Chapter 1, before turning to an analysis of the expansion of health information and advice catering for young women in Chapter 2. Chapter 3 examines debates on exercise for girls, notably the case study of cycling, and Chapters 4 and 5 focus on agencies involved closely in the moderation and promotion of girls' health around the turn of the twentieth century: girls' schools, the workplace and girls' clubs. During the last quarter of the nineteenth century young women became the targets not of anything as grand or unified as 'health campaigns', but certainly advice directed towards improving their health and wellbeing. This would equip them for a wide range of roles and in the best-case scenario produce the 'double gain' of smarter, fitter girl citizens who would subsequently become better wives and mothers. Though the state intervened in some arenas, notably by means of factory legislation addressing the hours of labour and conditions of work for women and girls and via the school medical service, the lead in addressing girls' health issues was taken largely outside traditional institutional settings and sites of medical practice, by voluntary organisations, schools, clubs and societies.[60] The Young Women's Christian Association, for example, was anxious not only to improve girls' working lives and to curb their potentially immoral behaviour, but also to balance work with useful recreation and continuing education, and to offer constructive advice on health, hygiene and nutrition. Many years before health began to be promoted in government health education campaigns and by organisations such as the New Health Society, good health was linked to new lifestyle agendas for girls, building on the fervent interest that had emerged in mid-Victorian Britain on health, sport and exercise for men and boys.[61] Though the advice offered was often framed in sentimental or patronising ways, or was even downright insulting in terms of the limits it suggested placing on girls' activities, much of it was innovative in its approach to health improvement and provided reams of information and advice on how to become fitter, leaner, stronger and more active to young women facing new challenges and opportunities. It also, while deflecting concern about biological limitations, placed growing emphasis on how girls' behaviour and responsible actions could govern – and possibly imperil – their health and prospects.

1
Unstable Adolescence: Medicine and the 'Perils of Puberty' in Late Victorian and Edwardian Britain

The late Victorian and Edwardian eras were in many ways times of great opportunity for young women, marked by increased access to secondary, college and university education and paid employment in factories, workshops, offices and professions such as teaching and medicine. New, less restrictive modes of dress and models of body image made headway, associated with the opening up a variety of exercise regimes, sports and recreations to adolescent girls. These opportunities paralleled and, to a certain extent, drove forward a process whereby established ideas of female weakness based on biological vulnerability were challenged and substituted by the ideal of the strong, fit and active modern girl.[1] This was a process that varied greatly across the social classes and, though working-class girls had far fewer opportunities in terms of employment or recreational outlets, as subsequent chapters will demonstrate, it potentially involved all girls regardless of status or location. None of these developments, however, went undisputed on medical, social or moral grounds, or all three combined. The emergence of the vibrant and ambitious modern girl triggered negative responses by social commentators eager to preserve the status quo in terms of the ideals of girlhood embodying the characteristics of femininity, docility and homeliness. These ideals were reinforced by medical theories which described young women as biologically unstable. This instability, many argued, was likely to be intensified by girls' inappropriate and harmful actions, notably their pursuit of new educational or emancipatory goals, which jeopardised their mental and bodily health.

This and subsequent chapters, however, will argue that depictions of girls as 'eternally wounded', frail victims of their biological weakness, were far from monolithic or impermeable. Yet they were certainly durable in many ways, spurred on by influential and widely read authors, notably the French historian Jules Michelet, who explained that for 15 or 20 days out of 28, women were indeed wounded invalids.[2] The late nineteenth century was notable for the production of a medical literature which dwelt long and hard on the diseases and disorders to which women were prone in connection with the reproductive process, the female life cycle more broadly and the rite of puberty in particular. Concerns about youth focused on boys and girls, but girls, having the extra instability of female adolescence and the onset of menstruation to contend with, were deemed particularly liable to disease, disorder and deviance in behaviour as well as bodily state. Anxiety about the proneness of girls to hysteria reached a peak during this period, while anorexia nervosa was labelled in 1873 as a disorder to which young women were acutely susceptible.

Negotiating the engagement of medical authors with the issue of girls' health in late Victorian and Edwardian Britain, this chapter will suggest that this was complex, fluid and sometimes contradictory, much more than an extension of arguments about women's limited physical and mental capacity and the centrality of reproduction, which Vertinsky has argued 'increasingly defined medical views of women's health and the productive boundaries of their lives'.[3] Anxieties about girls' intrinsic weakness and the dangers of passing through puberty were paramount in much of this literature. That much is clear. Yet, at the same time, this material indicated ways in which girls' health in general and puberty in particular might be managed to reduce its perils, in a form of damage limitation that would also enable girls to engage with the new opportunities opening up to them. Mary Lynn Stewart has pointed out that in late nineteenth-century France puberty was treated not necessarily as pathological, but as 'ambivalent', 'a transitional time when the body was susceptible to disease, and girls and women were prone to irrationality'.[4] And even doctors who have been singled out in the historical literature as holding particularly pessimistic views on the predestination of women and girls to poor health and the limits of their biological capabilities, appear to have been willing to explore the potential for girls to improve their health, albeit moderately and incrementally. Around the same time, debates about girls' mental and physical health began to spill over from medical arenas to engage a wider range of experts, social commentators and writers, as well as more diverse audiences, including girls themselves.[5] This literature tended not only to engage – sometimes critically – with the question of biological limitation, but also to promote approaches that would enable girls to enhance their health.

This chapter also examines the ways in which ideas about female adolescence as a distinct and vital phase of development became something of a national preoccupation at the turn of the twentieth century, speaking to concerns about the health and fitness of the nation's youth as a 'body' rather than the potentially poor health of individuals. Carol Dyhouse has suggested that historical literature on youth has focused strongly on boys, as has the literature passed down from this period, produced by social and welfare workers and club leaders.[6] With regard to the historical literature, this rings true; most studies have focused on boys.[7] But while young men were the main focus of many studies produced around 1900, a number also devoted sections – or even separate volumes – to the particular problems of adolescence for girls, including the health challenges associated with the onset of puberty and their long-term implications for motherhood and the security of the race and the nation.

Girls and fixed funds of energy

Dominating the late Victorian era and persisting well beyond it was the theory that the body contained only a limited supply of vital energy to fuel its physical and mental activities. This energy needed to be 'husbanded' and 'what was spent in one period was bound to be missed in another'.[8] For women, it needed to be preserved above all else for the reproductive process, and those arguing in support of an expanded range of opportunities for young women would have to contest this theory well into the twentieth century. 'Biological determinism', which promoted a limited sphere of action for women in terms of the space of the home and family and the function of childbirth and childrearing, had been expressed in some form or another from the eighteenth century onwards, but such ideas gained particular resonance by the mid-nineteenth century, spurred on by growing professional interest in the diseases of women, notably in the burgeoning fields of gynaecology and psychiatry.[9] An imposing array of historical literature has suggested the ways in which women were viewed increasingly as the products and prisoners of their reproductive systems which conditioned their health and social roles.[10] This was bolstered too by the more overtly political stance adopted by individual doctors who invested in ideas of gender difference, as women, challenging traditional roles and stereotypes, campaigned to enter public life and higher education and win the vote, as well as to pursue sporting and leisure activities which demanded physical effort and strength as well as a certain amount of 'nerve force', all at odds with dominant medico-social ways of viewing the female body. There were also strong economic incentives driving interest in women's bodies, as the new specialists dealing with female pathology sought audiences and markets for their practices and, to a certain extent, fuelled concerns about the risks associated with the female life cycle.[11] Much of this practice would focus on middle-class girls, and most medical interactions would take place in the homes of these girls rather than in medical institutions, though other institutional sites, particularly schools, also devoted a good deal of attention to girls' health and wellbeing and it should not be assumed that medical men had little or no contact with girls of the working classes. Certainly in their role as factory inspectors or through attendance at girls' clubs, they began to observe the impact of work on the physical and mental wellbeing of women and girls, while, via dispensary or hospital employment or in private practices devoted to poor patients, they were able to note the ways in which poverty imposed itself on girls' development and susceptibility to illness.

By the third quarter of the nineteenth century medical practitioners had evolved an 'economic' model to explain how women's reproductive systems interacted with other parts of the body, especially the brain, depicting 'the body as a closed system in which organs and mental faculties competed for a finite supply of physical or mental energy; thus stimulation or depletion in one organ resulted in exhaustion or excitation in another part of the body'.[12] The notion that women had fixed amounts of vital energy to draw on had particular resonance when it came to describing puberty and the onset of menstruation. The philosopher, political theorist and 'supreme ideologue of the Victorian period' Herbert Spencer famously described nature as 'a strict accountant' and objected to the 'forcing system' of girls' education which resulted in London drawing rooms full of 'pale, angular, flat-chested young ladies'. 'By subjecting their daughters to this high-pressure system, parents frequently ruin their prospects in life. Besides inflicting on them enfeebled health, with all its pains and disabilities and gloom, they not infrequently doom them to celibacy.'[13]

The two high priests of the 'fixed fund of energy' theory, Harvard Professor Edward H. Clarke and eminent British psychiatrist Dr Henry Maudsley, were remarkably successful in propagating their ideas on both sides of the Atlantic, reflecting a broader flow of medical information and medical texts between the two nations. In many ways, they were not presenting new ideas, but 'merely publicized the new scientific evidence for prejudices which had existed throughout the century'.[14]

Their pronouncements met with a receptive audience, particularly as they engaged with current and pressing issues, notably the question of whether young women were fit to attend at high schools and universities and the impact that brain work might have on their future capacity to bear healthy children. Edward Clarke argued that excessive physical or mental labour would reduce the supply of nerve energy to the female reproductive system, and that schoolgirls were at risk of developing neuralgia, uterine disease, hysteria and other damage to health and the nervous system.[15] If education interfered with the establishment of healthy, regular menstruation, according to Clarke, it could result in halted development, with girls graduating 'from school or college excellent scholars, but with underdeveloped ovaries. Later they married, and were sterile.'[16] Clarke did not deny that women were capable of learning, though he firmly opposed co-education, and stressed that 'they must do it all in woman's way, not in man's way'. 'For both sexes, there is no exception to the law, that their greatest power and largest attainment lie in the perfect development of their organization. If we would give our girls a fair chance... we must look after their complete development as women.'[17] As Julie-Marie Strange has suggested, Clarke grounded 'the notion that women's educational opportunities should be restricted in a language of *biology* and scientific empiricism rather than (fallible) social theory', thus setting 'an important precedent for shifting prejudice into the realm of physiological fact'; those disputing his claims, by trying to overcome their physiology, 'were, therefore, biologically and psychologically deviant'.[18]

  In many ways though, theories based on economics, biology and social theory conspired and, in the case of Henry Maudsley, also joined with concerns about social and mental degeneration. Maudsley warned in 'Sex in Mind and in Education', which appeared in an 1874 issue of the *Fortnightly Review* – an influential journal offering a platform for a wide range of social issues – that overspending vital energy threatened to cause menstrual disorders and mental breakdown, and potentially to destroy women's capacity to bear healthy children. Women, he argued, should not attempt to run alongside men, even women who did not aspire to marry and bear children: 'They cannot choose but to be women; cannot rebel successfully against the tyranny of their organization.' 'The important physiological change which takes place at puberty... may easily overstep its health limits, and pass into pathological change... nervous disorders of a minor kind, and even such serious disorders as chorea, epilepsy, insanity, are often connected with irregularities or suspension of these important functions.'[19] As Kate Flint has pointed out, Maudsley's voice henceforth 'had to be assimilated or countered by other commentators', and also significant was the fact that the debate he triggered 'spread from specialist texts to a range of publications with the potential to reach a wider readership'.[20]

Maudsley's article prompted an energetic and prompt response from Dr Elizabeth Garrett Anderson, published by the *Fortnightly Review* a month later. She robustly challenged the arguments he put forward directed at retarding girls' mental training, stressing that study was unlikely to weaken women's health, though teachers should protect girls from mental fatigue and violent physical activity. Education was, she argued, far less damaging than the idleness and boredom engendered by young women's traditional pursuits or, alternatively, the stimulus of reading novels, the theatre and the ballroom, late hours, vanity, frivolity and dissipation: 'There is no tonic in the pharmacopoeia to be compared with happiness, and happiness worth calling such is not known where the days drag along filled with make-believe occupations and dreary sham amusements.' Headmistresses and other reformers, Garrett Anderson maintained, had sought with marked success to improve, by emphasising exercise and hygienic practices, the physical development of girls alongside their mental training.[21] In some ways, Garrett Anderson's response can be framed as a straightforward expression of dissent on the part of a woman doctor who had striven long and hard to acquire her medical education in the face of fierce institutional, professional and personal opposition to women's uptake of medicine, particularly as she had strong affiliations to the feminist movement and campaigns to extend education to girls. Yet in other ways, Garrett Anderson shared common views and approaches to women's health with her male colleagues and, as an ambitious surgeon, regularly undertook gynaecological surgery, including the controversial and high-risk procedure of ovariotomy, which even many of her male colleagues roundly criticised.[22]

In any case, despite the protestations of Garrett Anderson and many other medical and social commentators, the impact of Clarke and Maudsley was durable. Girls were depicted as developing more rapidly than boys, thus using up their quota of energy more rapidly, which was further taxed by the demands of menstruation and subsequently reproduction. Dr William Withers Moore selected the theme of the limits of girls' resources of energy as the basis of his Presidential Address to the British Medical Association in 1886, arguing that education for girls was 'physiologically expensive' and detrimental to the good of the race and nation.[23] Moore contended that the years preceding puberty were as crucial as the years of puberty themselves: 'At no epoch of life is the necessity for maintaining the balance between the construction and destruction of nervous energy greater than in the period immediately preceding adolescence.' Meanwhile, overpressure between the ages of 15 and 20 was, he declared, likely to lead to atrophy of the reproductive organs and loss of inclination to become a mother.[24] Towards the close of the century, predominant Edinburgh alienist and medical author, Dr Thomas Clouston predicted a gloomy future; 'if the education of civilised young women should become what some educationalists would wish to make it, all the brain energy would be used up in cramming a knowledge of the sciences, and there would be none left at all for trophic and reproductive purposes'.[25]

The 'perils of puberty'

Incorporating, supplementing and rejuvenating the fixed fund of energy thesis, medical writing engaged vigorously in debates about puberty and its associated risks, producing a lively, and at times lurid, literature. In 1880 Dr J. Mortimer-Granville, whose interests included insanity and hysteria, warned in no uncertain terms about the dangers of the 'girl-youth', relating a grim scenario for parents and guardians who failed to manage this difficult epoch as girls approached maturity.[26] Aside from the dangerous influence of coarse, lascivious and depraved servants, and fathers who could not resist romping inappropriately with their young daughters and dangling them on their knees, Mortimer-Granville warned parents to watch for indications of illness, disease or special weakness, especially if there was a hereditary taint in the family. He described girls as being prone to palpitations, hysterics, fainting fits, giddiness and blood disorders, as well as outbursts of temper, moodiness, nocturnal tooth grinding, nightmares and sleepwalking. A hysterical cough could signal consumption, growing pains could indicate incipient rheumatism, or stooping a sign of deformity of the spine and limbs.[27] Mortimer-Granville's 'awkward age' could be short or prolonged, marked by a 'vague expression of weariness and ungainly stupidity or sullenness'; 'a courteous demeanour may conceal an ungracious heart: an attractive and innocent surface-character made up of smiles and pleasing manners, may veil a repellent and vicious nature'.[28] Mortimer-Granville signalled serious problems associated with puberty amongst boys as well as girls. Both sexes were liable to suffer as a result of over study, though girls were particularly 'over-educated in these mad days'.[29] Boys, according to Mortimer-Granville, were corruptible and likely to go astray, lazy, untruthful, moody, 'girlish' or weak; but, as girls faced the particular challenge of menstruation at puberty, they were described as excessively vulnerable and rather sinister. Boys were prone to a more curtailed list of illnesses and needed training and honing to develop their positive traits of energy and enterprise; physical lassitude, for instance, was to be treated, not by coddling, but by turning them out to paddle in brooks, or to work in the garden or on a farm.[30]

Few writers matched Mortimer-Granville's graphic rhetoric, but, given the risks of puberty for girls, many physicians urged careful watching on the part of parents for signs of ill health. This was imbued with additional urgency once it was established that menstruation was likely to occur at an earlier age in many young women, a factor associated with the enhanced development of the nervous system in wealthy girls, mental excitement and urban life.[31] Menstruation – though poorly understood in the late nineteenth century – was described as an absolute turning point in medical and social terms, marking the end of childhood, and in many ways the end of a healthy outdoors existence where girls played with their brothers, sharing their activities and physical freedom. Much literature focused on delaying the onset of menstruation through the prohibition of hot baths, spicy food, over-heated rooms, the excitement of dances or the theatre, novel-reading and late hours, substituting these with cold bathing, cool, airy rooms and a nursery diet.[32] Edward Tilt, Physician to the Farringdon General Dispensary and Lying-In Charity and Paddington Free Dispensary for the Diseases of Women and Children, cautioned in his 1851 textbook on the diseases of women not to bring girls forward, as 'to bring them to the full perfection of womanhood, is *to retard as much as possible the appearance of first menstruation'*. This retardation would be achieved by instituting a regime of

> rational food, rational hours of rest and of rising, and rational exercise at judicious times...the absence of sofas to lounge on – the absence of novels fraught with harrowing interest; it means the absence of laborious gaiety, of theatres, and of operas – the absence of intimacies which are of too absorbing a nature, and wholesome subjection of every minute to rule and discipline.[33]

Operas were a particularly 'potent engine of mischief', combining music, dancing, theatre and scenery, all laced with 'intense emotions', and were to be avoided at all cost.[34] George Black's advice manual, *The Young Wife's Advice Book*, first published in 1880, described how 'a lazy, listless life; undue mental excitement, either caused by the reading of sensational novels, by conversation or the like; late hours, irregular habits of sleep, highly seasoned articles of diet, and stimulants, have all a tendency to accelerate the occurrence of menstruation'.[35] Black also commented – bringing a positive, though also foreboding, tone to his account – that around the time of puberty a girl's demeanour altered, her bearing became more dignified, she became more retiring and 'there had begun to dawn upon her mind the consciousness of that important mission she is destined to fulfil'.[36] Alongside his warnings about operas, novels and sofas, Tilt, perhaps surprisingly given the tone of much of his writing, insisted that mothers should never let a girl be taken unawares at the onset of puberty, or keep their daughters on shortened allowances of food or limit outdoor exercise to them, which he referred to as 'bedlamite conduct'. He also urged a rounded education for girls and criticised the lack of attention given to health and exercise at girls' schools.[37] The debilitation resulting from novels and ballrooms was also referred to by Elizabeth Garrett Anderson. In her case, however, this exemplified the draining pointlessness of many girls' lives and led her to argue for an extension of their opportunities, which would lead to the abandonment of such trivial activities.

The majority of medical writers, however, saw the onset of puberty as the occasion to sharply apply the brakes to girls' activities outside of the home.[38] In 1891 *The Ladies' Physician* suggested that puberty made 'great demands upon the constitution' and that young women were particularly liable to disturbances of the physiological processes and to attacks of disease. 'Marked care', it was concluded, 'should be taken of the young girl during this period'.[39] Dr Robert Reid Rentoul had no reservations in marking puberty out as a time of unprecedented risk for girls in his 1890 pamphlet *The Dignity of Woman's Health*. A fervent eugenicist, Rentoul advocated the compulsory sterilisation of the 'unfit' and the mentally ill, as well as the prohibition of factory work to married women.[40] He described girls as fragile beings and the passage into womanhood as a time of grave danger:

a girl when becoming a woman should not have any mental or bodily labour to perform. She should therefore neither study nor work, but have only good food, exercise, sleep, and clothing. If she accomplishes the beginning of this important phase of her life with vigour and success, she will have secured one of the chief ends of her existence.[41]

According to Rentoul, this chief end was motherhood. His 1890 pamphlet was dedicated 'to mothers – present and future' and he took the ovary as being emblematic of the duties of girls passing into womanhood:

> Show an uninterested observer an ovary. To such it appears only a bit of flesh. Yet on two such little organs – weighing about eighty grains each, and having a length of an inch and a half – depend not only the future of the world's population, but also the health and happiness of their owner.[42]

Before the onset of menstruation, Rentoul suggested that girls could and should engage in healthful activities, running, rowing, skipping, fencing, walking, bathing, tennis, horseback riding and so on. At puberty these should cease or at least be severely curtailed: 'I go so far as to say, that if a girl when at school shows any signs of approaching womanhood she should be taken home. What is the use of sacrificing health for the sake of "grinding up" a little history, geography or music?'[43] During the first five to six months when menstruation was being established, Rentoul urged that girls must stay at home and rest on the sofa 'until the system becomes used to the new condition...The effort to develop into womanhood requires all the best energies and strength a girl can possess.'[44] Dire consequences, he argued, would follow if these recommendations were not obeyed, including weakness, disability and even death.[45] Though Rentoul's eugenicist views were deeply contentious and controversial, his stance on the need for girls to rest, to suspend rigorous activities and await the blossoming of womanhood was shared by many medical writers.

John Thorburn, Professor of Obstetric Medicine at Victoria University, Manchester, argued that girls were capable of steady work as they passed through puberty, but, with regard to higher education, they should 'temper zeal with discretion'. At the onset of menstruation, which coincided with the take-up of secondary education, 'an entirely new element comes into play...mothers must be taught that their daughters cannot always with impunity undertake the same continuous mental strain, or the same hours of close attention which may be harmless at non-menstrual periods'.[46] Pointing to yet more severe risks, girls were described by Henry Maudsley as more liable to pubescent insanity than boys, particularly if there were indications of predisposition to mental illness. The onset of menstruation marked a huge turning point in their development:

> The great internal disturbance produced in young girls at the time of puberty is well known to be an occasional cause of strange morbid feeling and extraordinary acts...irregularities of menstruation, always apt enough to disturb the mental equilibrium, may give rise to an outbreak of mania, or to extreme moral perversion.[47]

Medical texts warned that menstruation – risky enough when normal – was likely to involve excessive bleeding and pain, be irregular or cease altogether, and a large number of pages in medical textbooks devoted themselves to these concerns. Chlorosis, or green sickness, was related to suppressed or irregular menstruation in young women and a host of other symptoms, including debility, breathlessness and palpitations, odd dietary habits and a distinctive green tinge to the skin. As Loudon has suggested 'those physicians whose experience lay wholly in private practice could believe in the essential connection between chlorosis and feminine purity and delicacy among the affluent; those with appointments at hospitals and dispensaries could not'. It was reputed to afflict working-class girls living in confined urban conditions, following sedentary occupations, as well as high-ranking young ladies; boarding schools in particular were seen as breeding grounds of chlorosis.[48]

Sir Andrew Clark, teacher of clinical medicine at the London Hospital, acknowledged that chlorosis occurred amongst girls of all classes. He related it to 'violations of physiological laws common to every case' and also to girls' wilful and perverse behaviour.

> In the period between the advent of menstruation and the consummation of womanhood there arise physical, mental, and moral changes which greatly influence the girl's habits of life and thought...She thinks of her appearance and tightens her waist. Afraid of getting fat, she stints herself in food, and eats of only dainty things. With her sense of modesty deepened, she is shy of being seen about the closet. Unprompted by nature, and perhaps disdainful of such affairs, she omits the daily solicitation of the bowels.[49]

Consequently, girls' bowels became obstructed, which led to damaging substances being absorbed into the blood, producing anaemia. Clark recommended doses of iron and evacuation of the bowels with laxatives to relieve the condition. He also encouraged adjustments in lifestyle, attention to regimen and the need to develop a sense of purpose to bridge the gap between girlhood and womanhood:

> On rising, take a tepid sponge bath; dry quickly, and follow with a brisk towelling. Clothe warmly and loosely... Have four simple, but liberal meals daily... Walk at least half an hour twice daily, and as much more as strength and convenience will permit. Retire to bed about ten, and repeat the sponging and towelling. See that your bedroom is cool and well ventilated. Lead a simple, regular, active, occupied, purposive life; and do not notice or distrust yourself.[50]

The title of Clark's article pointed to the occurrence of chlorosis 'between the Advent of Menstruation and the Consummation of Womanhood', thus referencing the long-held view that chlorosis was related to a state of suspended reproductive activity in unmarried young women, which only marriage and motherhood would permanently resolve.[51]

By the 1890s the connection between chlorosis and a low red blood cell count had been established and thereafter iron was recommended as a treatment. Alfred Lewis Galabin, Lecturer on the Diseases of Women at Guy's Hospital, London, warned that the commencement of menstruation was likely to trigger chlorosis, 'the extra demand which thus arises having proved too much for the feeble powers of the system'. However, girls were likely to recover if they were treated with a combination of iron and careful hygienic management, diet, fresh air and 'judiciously regulated exercise (the most effectual form of which is riding on horseback), cold fresh, or still better, salt water baths, and change of air and scene'.[52] John Thorburn maintained that chlorosis and anaemia in girls was triggered by poor diet, a lack of exercise, light and air and by 'whatever tends to lower the nervous tone, in the way of mental anxiety, loss of rest, over-exertion, – especially mental, abnormal sexual excitement, or the various troubles which the adolescent girl experiences at home, at school, or in society'. He prescribed iron or arsenic and salts of manganese, as well as sun-ray therapy and the rest cure; the latter was deemed particularly useful when the condition was linked to a 'nervous affection'.[53]

It was not just orthodox medical practitioners, gynaecologists and psychiatrists who expressed concern about the passage of girls through puberty. Edward Johnson, a qualified doctor who converted to hydropathy in his despair at orthodox medicine's approaches to female disorders, described numerous mishaps connected to puberty and menstruation in a volume devoted to female ailments. While many writers on the subject were anxious to stay the onset of menstruation, Johnson was far more concerned about the potential impact of delayed puberty and the 'positive disease' which could follow or from this.

> She will become dull, sad, inactive, languid, and listless; or sullen and perverse, and fond of being alone. Her appetite will either leave her or become morbid, and she will acquire a taste for unwholesome food, or for all sorts of trash, such as chalk, slate pencil, vinegar, and sometimes even dirt...The bowels will become constipated, the breath offensive; very slight exertion will fatigue her, and hurry her breathing.[54]

While the 'systematic drugging' of allopathic approaches 'shatters nerves, poisons their blood, wrecks the health', according to Johnson, hydropathic treatments – wash downs, sitz baths, wet sheet wrapping, rest, plentiful and plain diet, and ample exercise – would steer girls successfully through this 'momentous epoch'.[55] Homeopathist Edward Ruddock referred in graphic terms to the blossoming of girls at puberty, as they veered off in developmental terms from boys, with their full and elegant figures, the pelvis taking on a sexual character, breasts rounded, and 'so securing one great object for which the female was created – the reproduction of the species'.[56] His advice on treatment was similar to that of orthodox medical authors who advised staying the onset of menstruation – avoid hot baths, stimulating food and drinks, overheated rooms, dancing, novels and late hours, as 'such habits and indulgences tend to occasion precocious, frequent, copious, or irregular menstruation'.[57] Mothers were advised by Ruddock to be vigilant as menstruation became established, 'to keep an account of dates and other particulars, and prevent all unusual exposure for a few days before the expected flow, such as night air, damp linen, thin dresses, wet feet, balls, and evening entertainments'.[58] 'Carelessness, or constitutional delicacy, may render this period extremely dangerous by the propagation of new forms of disease, or the development of any latent germs of disorder which have existed from birth'.[59]

Of all the afflictions that could strike down adolescent girls, hysteria was the most striking and a rich source of copy in medical and popular literature.[60] It was ascribed alarmingly by Walter Johnson to 'Polite Education', the grand cause of hysteria – that which puts out the eyes and lames the limbs, and distorts the features of the young and beautiful; that which prompts the canine bark, obstructs the breath, and wrings the brow with anguish; that which melts the women of England into powerless babes, ... deforms the moral beauty of their souls, and shatters their intellect.[61]

Dreadful manifestations, involving muscular contractions, weakness, fits, paralysis, squints, tics, pain, defects of speech and loss of memory were reported in medical journals. Even doctors who regularly reported such cases, however, struggled to define hysteria and work out its relationship with disorders of the mind and brain or organic disturbance, though it was repeatedly linked to girls who were defined as 'emotional', as well as to over-education, mental strain, luxury, unhappy circumstances and shock.[62]

The prominent gynaecologist and authority on nerve disorders, Dr William Smout Playfair described a case resulting from the 'evil effects of over much education and mental strain in a clever girl of highly developed nervous organisation'. She had broken down at the age of 14 and for four years did not leave her bed or move her lower limbs, refused all food except milk, oranges and biscuits, and had a loud barking cough.[63] Psychiatrist F.C. Skey affirmed that hysteria could occur in older women, but was

most prevalent in the young female members of the higher and middle classes, of such as live a life of ease and luxury, those who have limited responsibilities in life, of no compelled occupation, and who have both time and inclination to indulge in the world's pleasures – persons easily excited to mental emotion, of sensitive feeling, often delicate and refined.[64]

Such women were not necessarily weak minded, Skey added: 'It will often select for its victim a female member of a family exhibiting more than usual force and decision of character, of strong resolution, fearless of danger, bold riders, having plenty of what is termed *nerve*.'[65] Dr Horatio Bryan Donkin, neurologist, progressive thinker and doctor to Karl Marx and the feminist Olive Schreiner, provided a complex reading of the condition, though he also ascribed it to the extreme selfishness of its victims. In an 1892 essay on 'Hysteria', Donkin confirmed that hysteria was most likely to afflict young women subject to nervous imbalance triggered by enormous bodily change, particularly in their sexual organs, but also explained how social conditioning and the frustrations it caused were of great relevance:

> The nervous balance is thus in especially unstable equilibrium. With this greater internal stress on the nervous organism there are in the surroundings and general training of most girls many hindrances to the retention or restoration of due stability and but few channels of outlet for her new activities. It is not only in the educational repression and ignorance as regards sexual matters of which she is the subject that this difference is manifest, but all kinds of other barriers to the free play of her powers are set up by ordinary social and ethical customs. 'Thou shall not' meets a girl at almost every turn. The exceptions to this rule are found in those instances of good education or necessity, or both, have regular work and definite pursuits.[66]

This compared, Donkin explained, with conditions for the young man, 'whose comparative freedom and the various and necessary occupations...offer many safety valves for his comparatively minor nervous tension'.[67] Maudsley too framed female adolescence as deeply problematic and suggested – setting up something of a tension with his well-known pronouncements on the adverse effects of higher education for young women – that girls suffered more than youths, not just because of the influence of their reproductive organs on mind and the challenge of menstruation, but also because 'the range of activity of women is so limited, and their available paths of work in life so few...that they have not like men, vicarious outlets for feeling in a variety of aims and health pursuits'.[68] Yet he also rounded on hysterical young women who 'believing or pretending that they cannot stand or walk, lie in bed or on a couch all day...objects of attentive sympathy on the part of their anxious relatives, when all the while their only paralysis is a paralysis of will'.[69]

More considered interpretations of the propensity of young women to hysteria referred directly to the challenges imposed by society and culture which co-existed or even trumped biological weakness as causal factors. Many doctors also denied the prevalence of 'shamming' by girls eager to obtain attention by means of their prostration. Yet hysteria retained its association with less than desirable traits: boredom, underemployment, excessive pampering, selfish and vindictive behaviour, and the sexual excitement linked to puberty. Playfair described hysterics as 'thoroughly enjoying their life of inert self-indulgence'.[70] Girls were in a no-win situation: inactivity and lack of stimulation was bad, but so too, according to some medical authorities, was mental overstrain and exertion.

Neurasthenia, a 'more prestigious and attractive form of female nervousness than hysteria', associated with sensitivity, education, high aspirations and good breeding, in contrast to grasping, selfish and disruptive hysterics, was characterised too by loosely defined symptoms, including fatigue and prostration, sleeplessness, spinal tenderness and neuralgia, hypochondria, dyspepsia and constipation, loss of appetite and voluntary starvation.[71] Treatment by means of the rest cure, pioneered in the US by Silas Weir Mitchell and imported to Britain by Playfair, emphasised seclusion and rest, diet and regimen, supervised ideally by strict medical attendants, as well as the absence of stimulation of any kind, including reading, writing or other creative activities. However vaguely defined, neurasthenia claimed as its victims young women and women in transition from domestic to professional roles, Playfair referring, as he also would in cases of anorexia, to 'overwork in the modern system of higher-class education in girls, whose physical health is unfitted for the efforts they are unwisely encouraged to make' as the cause.[72] Novelists in England between the 1890s and the First World War closely associated nerve disorders with highly strung modern women and high achievers, and fiction referring to the phenomenon of the New Woman, as well as describing her discord with society and its limitations for women, focussed on nerves, disease and death, a consequence of the disjuncture between the New Woman's ideals and her mental and physical capabilities and environment.[73]

Anorexia nervosa, defined simultaneously in 1873 by Gull in Britain and Lasègue in France, was, like hysteria, associated with selfish girls, who destroyed family life, as their wasting bodies became the centre of concern.[74] Charles Lasègue described the typical age of onset as between 15 and 20 and related it to inappropriate romantic expectations, blocked educational or social opportunities, or struggles with parents, while Gull suggested 'subjects of this affection are mostly of the female sex, and chiefly between the ages of 16 and 23'.[75] Like hysteria, such cases attracted a great deal of interest in the medical press, often accompanied with images of emaciated young women contrasted with their recovered selves following the physician's intervention. In his 1874 publication 'Anorexia Nervosa', Gull presented cases of three young women referred to him by general practitioners.[76] Compared with Lasègue who highlighted the psychological aspects of the condition, elaborating on the mental state of the patient and her family, Gull tended to emphasise the medical aspects of his cases, extreme thinness, loss of appetite and amenorrhea. Yet he also referred to his patients as 'wilful' and, in the case of 15-year old Miss C, 'obstinate' and 'restless', her 'mind weakened':

The great difficulty was to keep her quiet, and to make her eat and drink. Every step had to be fought. She was most loquacious and obstinate, anxious to overdo herself bodily and mentally... Rest, and food, and stimulants as prescribed, undoubtedly did her a great deal of good. She used to be a nice, plump, good-natured little girl. Believe me, &c.[77]

Joan Jacobs Brumberg has suggested that Gull's 'before' images emphasised a look of derangement – 'Miss C was unquestionably emaciated, and her accentuated jawline and profile were reminiscent of a willful young horse.' The 'after' photograph emphasised tranquillity and ordinariness, and 'assumed the demeanor proper to young women of their role and station and lost the look of dour petulance that Gull believed characteristic of the anorectic'.[78] The description of this case emphasised the struggle to assert authority over the patient and to 'rein in' unfeminine and inappropriate behaviour (see Figure 1.1).

In 1888 Gull reported the case of Miss K.R. to the *Lancet*. Aged 14, she had abruptly stopped eating and declared a repugnance to food. Gull declared the case to be one of perversion of the 'ego' and marked by 'the persistent wish to be on the move, though the emaciation was so great and the nutritive function at an extreme ebb'. A nurse was obtained from Guy's Hospital to take close care of the girl and to impose a regime whereby she took light food every few hours. She recovered within a couple of months.[79] William Playfair noted that alongside food refusal, the condition was marked by extreme restlessness and suppression of menstruation, and he attributed it largely to overstrain; he had seen 'many instances in young girls which have followed severe study for some of the higher examinations for women now so much in vogue'. Other causes, according to Playfair, included bereavement, money losses, disappointment in love and the strain resulting from over-athletic activity.[80] Treatment of anorexia nervosa typically involved removal of the patient from corrupting environments, principally the home, hospitalisation in some cases or the employment of a nurse, regular feeding with small meals, and, if necessary, force-feeding and enforced rest, approaches which emphasised the infantilisation of young women.

32

Figure 1.1   Two 'before and after' portraits of a young girl, Miss C., with the condition anorexia nervosa.

Source: From William Gull, 'Anorexia Nervosa (Apepsia Hysterica, Anorexia Hysteria)', *Transactions of the Clinical Society of London*, 7 (1874), 22–8, Case C, facing p. 26 (Wellcome Library, London).

Elaine Showalter has illustrated how 'approved medical attitudes towards anorexic girls' (if not standard treatment regimes) were described by the New Woman writer Sarah Grand. In her novel *The Heavenly Twins*, published in 1893, the fictional London nerve specialist Sir Shadwell Rock sent the anorectic girl away from her family to be treated by a hard, cold stranger: 'When she fainted she was left just where she fell to recover as best she could, and when any particular food disagreed with her, it was served to her incessantly.' The girl finally confessed that she had been 'shamming from beginning to end'.[81] As with hysteria there was a great deal of debate about the aetiology, symptoms and causes of anorexia, but it was strongly associated with bad parenting and wilful girls. Doctors condemned families who spoilt their daughters, were over-indulgent and lacked moral authority, and girls were rebuked for their selfish behaviour, including an urgency to study and take on active sports, which triggered their disgust of food.

The behaviour of young girls, such as the craze for the tight-lacing of corsets, only served to exacerbate the biological disadvantages of their unstable young bodies and complete, as Jalland and Hooper have put it, their 'psychological devaluation'.[82] Described as silly, perverse and vain, tight-lacing was castigated in medical literature and popular media, for inducing unnatural excitement and vice, as well as severe organic disturbance, deformity and, in extreme cases, death or serious disease such as cancer.[83] Corsets, one female author opined, retarded the function of the heart, lungs, liver, stomach and bowels, prevented full development of the breasts and laid 'the foundation for future troubles when the girl shall become a woman and mother'.[84] Their wearing entailed the pressing down of the internal organs, resulting in 'a long train of evils, known as "female complaints" which unmarried women ought to be ashamed to confess to, but which are frightfully prevalent among them'.[85] Howard Kelly referred to the attire that women adopted at puberty, driven by 'the awakened desire of making herself attractive', as 'the despair of the hygienist'. John Thorburn, meanwhile, described 'the tying of a "dry goods store" round the waist or abdomen, in the form of heavy petticoats' as 'highly injurious' and corsets with steel or whalebone 'as absurd for a healthy woman as steel-jointed leg pieces would be for a healthy man'.[86] Surgeon Frederick Treves summed up five damaging effects on health resulting from the corset – to the viscera and internal organs; respiration; circulation and the heart's action; the muscular apparatus of the trunk, and the general outline of the body.[87] One particular risk for growing girls was that the function of the muscles was 'absorbed by the corset' and resulting muscular weakness was likely to lead to curvature and other deformities of the back.[88] Yet girls, it was regretted, persisted in wearing them, flouting medical pronouncements on the associated dangers to motherhood, physique and health. It was only as sport and exercise became more common for girls – and as fashions changed – that tight-lacing was abandoned.[89]

Though associated primarily with boys dissipating their vital energy, a number of leading members of the British Gynaecological Society claimed that masturbation was also widespread among young women and was a habit often acquired at school.[90] The influential gynaecologist Robert Lawson Tait argued that its most pernicious effects were met with when the contamination reached a congregation of young women, as in a girls' school', but nuanced this by concluding that the effects of masturbation had been greatly overrated, that it was rare amongst girls, and that most girls could be induced to give up the practice once the risks had been pointed out.[91] Nonetheless, masturbation was associated with menstrual disorders and hysterical symptoms, and for a small number of girls and women was treated by the horrific procedure of clitoridectomy.[92] The heated controversy on clitoridectomy, a preoccupation of a number of medical journals, including the *Lancet*, was notable for its reflection of the diversity of views on female physiology and sexuality, but highlighted for many doctors a dark and threatening aspect of female sexuality divorced from reproduction.

Images of girls' susceptibility to mental and physical meltdown were durable and the notion of the female body as intrinsically sick and dominated by its reproductive organs and functions was often, as Jane Wood has pointed out, remarkably crude compared with 'the profound treatises written by the same doctors in other areas of neurophysiology'.[93] As late as 1913 Dr MacNaughton-Jones was still explaining how at puberty 'Every nervous ill and weakness is accentuated by masturbation ... the morbid imaginative impulses are excited with consequent fits of mental depression and sense of fatigue – and the erotic girl merges into neurasthenic womanhood.'[94] In 1908 the American gynaecologist Howard A. Kelly explained that great care should be taken not to attract the attention of girls to their sexual organs nor to sexual things; puberty was to be feared with the exciting forces of the 'monthly rush of the blood to the genitals, the friction of the napkin, the suggestiveness of the hot-water bag, the lying awake in day dreams'.[95] Kelly did, however, advocate a proactive response to counter this danger: an outdoor life and varied interests, nature study, swimming, hydrotherapy, gymnastic games, skating, tennis, golf and horseback riding and a nutritious diet.[96] Substantial sections of his book *Medical Gynecology*, which was also published in England, were devoted to the hygiene of puberty and the hygiene of the schoolgirl, and much of this centred on practical, commonsense advice, emphasising the importance of proper nutrition, open air exercise, sleep, sensible clothing and cleanliness during menstruation.

Adolescence and girlhood at the turn of the century

The relationship between girlhood and health was tackled in new ways as adolescence became a focus of heightened attention around the turn of the twentieth century.[97] Far from ideas about girls' fixed funds of energy fading into insignificance at this time, it gave such notions a new rationale and significance, and it has been argued that this period saw an expansion of biologically deterministic views, the early 1900s being characterised by both eugenic thinking and 'a regression to the more extreme notions of female inferiority and sex differences that had distinguished the 1860s and 70s'.[98] Adolescence was deemed a unique phase of development and a time of opportunity as well as challenges, but for girls these challenges were depicted as potentially threatening their health, stability and even sanity, while for boys the expectation was that these challenges would be robustly responded to and overcome *en route* to manhood. There was a proliferation of studies on adolescence during the Edwardian period, and concerns about the nation's youth were amplified during the South African (Anglo-Boer) War and the First World War as they dovetailed with anxieties about the falling birth rate (particularly amongst the middle and upper classes) and the threat of race degeneration, persistent poverty and high infant mortality.[99] At the very time when 37 per cent of British volunteers for the South African War were being rejected as unfit for service, young women appeared to be turning their backs on motherhood and thus their duties as British citizens to pursue education, careers and personal fulfilment.[100]

These concerns were absorbed into medical writing, as well as the work of educationalists and social workers, particularly in the emerging field of child studies, and fuelled eugenic writing on the topic of adolescence. This new interest in youth was most strongly associated with the writings of the psychologist, educator and ardent eugenicist G. Stanley Hall, whose influential study, *Adolescence*, was published in 1904, and widely read in Britain and America.[101] According to Stanley Hall, male adolescence was a period of ambition, growth and new possibilities, while for girls adolescence was predominantly governed by instability and the need for special protection by parents and others involved in their care.[102] Carol Dyhouse has discussed in some detail the florid, cloying and sometimes odious prose which typified Stanley Hall's writing on girls.[103] His work constituted a backlash against feminist endeavours and girls' education in particular. Girls should be told early, he argued, that the accent will lie chiefly on childbearing: 'In this way the girl will be anchored betimes to what is really the essential thing, viz., reproduction and the carrying beneath her heart and then bearing children which are the hope of the world.'[104] He described young women as 'buds that should not blossom for some time', their behaviour marked by flirtatiousness, fads, weepiness, giggling, whimsicality, secretiveness and a strong dislike for study.[105] Puberty was particularly hazardous for young women, and Stanley Hall cautioned 'reasonable' girls to conserve their energy as menstruation was established; they should be urged to withdraw from society and school, and to enjoy the luxury of 'occasional slight illness and the indulgences it brings'.[106] He elaborated on an ideal regimen for girls approaching puberty. They would be sent to isolated institutions in the countryside, where their health would be carefully monitored, and attention devoted to diet, sleep, gentle outdoor exercise and dancing (especially the stately minuet) and manners.[107] 'Another principle should be to broaden by retarding; to keep the intuitions to the front; appeals to tact and taste should be incessant; a purely intellectual man is no doubt biologically a deformity, but a purely intellectual woman is far more so.'[108]

Others reflected Stanley Hall's concerns and arguments. Phyllis Blanchard, an associate of Stanley Hall, whose published work, like his, also had a broad impact in Britain, emphasised the dangers of sexuality associated with female adolescence, which marked the commencement of mental diseases such as hysteria and dementia praecox.[109] In a paper read to the Childhood Society in 1904, M.E. Findlay advocated rest for girls during puberty in order to avoid the maladies of anaemia, chlorosis and hysteria, and pointed out that

> Every girl has to pass during her birth into womanhood through a period of storm and stress; and, as every individual is endowed with only a limited amount of energy and momentum, its seems a matter of mathematical certainty that at a time when the more fundamental organs of the body and the deeper emotions are making special demands, the higher forms of brain and intellectual activity must withdraw theirs, otherwise the superstructure will eventually incur the danger of falling through the base being weakened.[110]

School, Findlay suggested, must be radically adapted to suit girls during the teenage years. Feminist eugenicist Arabella Kenéaly, meanwhile, envisaged girls' adolescence as a chrysalis stage – setting in around the age of 12 – through which girls had to pass before emerging as fully formed women:

> From having been a strong, young, active, boy-like creature, now – provided her development be allowed to take the normal course – the girl loses physical activity and strength. A phase of invalidation sets in. Instinctively, she no longer runs and romps. New languors invest her in mind and in body. She is indisposed to brain-work or to much exertion. She lounges and muses. Her mind is clouded with the mists of awakening sensibilities. She suffers from lassitudes... She becomes a complex of disabilities, indeed; disabilities which indelicate, sickly or over-taxed girls show in chlorosis, anaemia, hysteria and other ills.[111]

J.W. Slaughter, who aimed to introduce Stanley Hall's theories to a wide audience of teachers and youth workers, adopted a more positive stance towards adolescent girls, though he believed that there were limits to what girls could achieve and that motherhood should remain their central objective. While condemning the kinds of girls' schools that focused simply on 'embellishment' and preparing young women for the marriage market, he also decried the newer schools and colleges for girls which simply duplicated the ethos and curriculum of boys' schools, threatening to 'do violence to the most important of all factors – woman's health and her own feminine nature'.[112] Dr T.N. Kelynack's National Health Manual on *Youth* struck a promising note when it turned to a discussion of girls, explaining that, for what he labelled the 'normal girl', 'adolescence comes practically unannounced'. For such girls, puberty proceeded in an orderly fashion, developmental changes were noted, and then forgotten. Girls could expect to be a little tired, to experience a little abdominal pain, but little else.[113] Unfortunately, however, according to Kelynack, the normal girl was an extremely unusual creature, and

> the placid picture of sane and wholesome adolescence... rarely met with. Only a small percentage of our girls to-day approach and pass the adolescent stage without disturbances of mind and body more or less grave. The pathology of the menstrual function alone fills many pages of our medical text-book, and how much of the surgical and

medical gynaecology of later years is the direct or indirect outcome
of the pathology of adolescence has yet to be determined.[114]

However, Kelynack's volume, one of a series of four manuals he edited
intended for social workers covering infancy, childhood, school life and
youth, also offered practical advice on how to mitigate this with warm
baths, rest, diet and sensible clothing.[115]

Around the years of the First World War the concept of eugenics
was spurred on by concerns about national efficiency and the avoidance of motherhood by the better classes 'shirking' their 'racial' duty.[116]
Prominent amongst these were a cluster of medically qualified, feminist women, notably Mary Scharlieb and Elizabeth Sloan Chesser, who
wrote extensively on motherhood and mothercraft, girls and adolescence, sexual hygiene and race regeneration, much of their work being
aimed at a general readership.[117] In terms of actual membership of the
Eugenics Society, established in 1908, the impact of eugenics appears to
have been limited, and Dorothy Porter has contended that public health
policy was little affected by the plans of eugenicists.[118] Yet, stretching
beyond its organisational confines, through women such as Scharlieb
and Chesser, eugenic ideas cascaded down to broad audiences through
textbooks for school children, advice literature on girls' health and public lectures. Mary Scharlieb, who was appointed gynaecologist at the
Royal Free Hospital in 1902 following medical work in India, believed
that women had the right to pursue professional careers. Yet she also
stressed the importance of motherhood for the national good and was
suspicious of birth control and the falling moral and Christian standards
it engendered.[119] She endorsed many of Stanley Hall's opinions, though
she challenged others, including his views on the irreligious nature of
young women. She described girls as fundamentally unstable; the 'per-
versity of adolescent girls make the proper care of their bodies extremely
difficult',[120] and warned that 'over-fatigue of any part [of the organism]
means the over-fatigue of the whole, and we have learnt that it is worse
than useless to expect that girls whose mental powers are over-taxed
can yet retain vigour and health of body'.[121] All those charged with
the care of girls, be they mothers, teachers or doctors, were enjoined
to instruct and caution girls 'in what is veritably an hour of need'; adolescence was seen above all as a period of flux and change, when girls
had but 'slight stability' to rely on.[122] However, Scharlieb saw the challenge of female adolescence largely as one of management, and given
careful attention to rest, diet, exercise and attire, girls had considerable
potential:

Such then is the adolescent girl, a bewildering mixture of the bright, the beautiful, the heroic, the difficult, the elusive, and the incoherent. The problem for parents and teachers is, how are we to emphasize and develop all that is desirable, and how are we to eliminate, or at any rate repress, the less desirable side of the girl's nature...although our patience, tact and wisdom are likely to be taxed to the uttermost, yet the reward is great and the possibilities are great.[123]

Elizabeth Sloan Chesser, one of the second generation of medical women to qualify in Britain, graduated in Glasgow in 1901. Known particularly for her journalistic enterprises and large number of publications on health issues and popular psychology, Chesser produced numerous books on women's and girls' physiology and health and motherhood, which reflected a more pronounced interest in class and differential fertility than Scharlieb, 'the direction of eugenic debate in the period leading up to the First World War'.[124] She advocated state welfare for the poor, arguing that mothers and children were a national resource, but also urged individual responsibility for health and fitness and the importance of motherhood to middle-class women and girls. Though preoccupied with the need to preserve the health of young women for future motherhood, Chesser urged girls to engage fully in civic life and meaningful employment until marriage, and her health advice focused on increasing capacity for taking on a variety of activities.

Conclusion

Medical texts produced a huge amount of commentary and extreme examples of the damaging effects of incorrectly managed puberty. Some of the conditions associated with puberty, such as anorexia nervosa, were indeed alarming and visibly shocking, as were fashions which strove for 18-inch waists, and feminist writers and headmistresses as well as physicians would despair of girls' persistence in wearing tight-laced corsets. Careful reading of such texts, however, also reveal complex, nuanced and ambiguous views on women's and girl's health issues on the part of medical authors and some pragmatic responses to the trials and tribulations of female adolescence. Though Howard Kelly, for example, included hair-raising remarks on puberty and menstruation in his publications, he also advocated straightforward approaches to its management. Henry Maudsley, best known for his pessimistic diatribes on degeneration and girls' capacity for ill health, gave out mixed messages on the problems of puberty for girls, and attributed ill health to under-activity and boredom as well as over-demand on vulnerable systems with limited supplies of energy. The same can be said for Playfair's pronouncements. In 1896 Playfair critiqued schools with little or no provision for exercise and for ignoring their duty of care in inquiring into the health of their pupils, who in some instances were urged to study for seven to eight hours a day. Girls, Playfair suggested, should be able to enjoy the advantages of secondary education, but their health needed to be observed and monitored, and a balance achieved between study and outdoor recreation.[125] Citing a leading article from the liberal review, the *Speaker*, Playfair also praised English girls for setting good examples of health, referring to 'Lawn Tennis Girl',

> an excellent example of the healthy, well-developed, and unsentimental girl – the girl who does not think it necessary to devote herself to the study of her own emotions, and who finds in active physical exercise an antidote to the morbid fancies which are too apt to creep into the mind of the idle and self-indulgent.[126]

Doctors, while strongly shaping topics and debates in the lay media on the medical challenges of girlhood, may also have been influenced not just by their professional colleagues, but by their own reading of lay publications, as well as the institutional links that they forged, particularly with girls' schools.[127] Meanwhile, as Nancy Theriot has suggested, female patients were potentially important contributors in doctor-patient interactions, who helped shape diagnoses and doctors' interpretations of disease.[128] The girls' voice is largely muted in the sources drawn upon in this chapter, but a number of physicians were convinced in the case of hysteria that girls were either shamming or using their disease to create and negotiate positions of influence in the household and beyond it, though others concluded it denoted frustration at their limited spheres of activity.[129]

A new wave of anxieties about the 'girl youth', embedded in wider concerns about the dangers of adolescence and publicised by authors such as G. Stanley Hall, followed hot on the heels of debates on the challenges posed by the New Woman in the mid-1890s. In the build up to the First World War concerns were again heightened about girls' ability and willingness to become mothers, with health at this point being defined more as an obligation for 'every girl, in duty to herself and to the race' rather than as a pathway to personal enhancement and fulfilment.[130] The impact of this new turn and enhanced focus on girls as citizens and race mothers will be returned to in later chapters and the conclusion, which also questions how far the First World War can be depicted as a time of reversal for modern girlhood.

For many medical authors at the turn of the century girls traversed adolescence on a knife edge; many forms of disturbance – mental and physical – were liable to tip them into ill health, weakness, invalidism, insanity, sexual depravity and even death, or imperil their ability to become mothers. Biological interpretations of women's bodies and health undoubtedly remained very significant, but alongside biology, environment – particularly new and challenging environments such as girls' schools – behaviour and emotional state were described as colluding to make the passage of girls through adolescence taxing and risky. Yet some medical practitioners moderated and nuanced dominant discourses, and puberty was framed as a time of transition as well as crisis. Many doctors argued that puberty simply required careful management, which led them to offer information and advice on regimen and life style to the broader public – themes which will be explored in the following chapter – and many were ambivalent, their views contradictory and inconclusive. What is important perhaps is not what was agreed or disputed but the fact that so many ideas and viewpoints circulated on female adolescence, its risks and its potential within and increasingly beyond medical literature.

It is worth also bearing in mind Michel Foucault's assertion that, as scientific discourse about the physical capabilities of the body increases, it tends to be brought increasingly within the orbit of professional control and power.[131] So while interpretations based on women's biological predestination might become more nuanced or even diminish, the role of many doctors shifted to providing information and advice to the lay public, providing a road map to guide girls through puberty and guidance on how to enhance health. So too, did the range of experts involved in this process expand to include headmistresses, club workers, gymnastics teachers and journalists. While this chapter has focused predominantly on medical literature, the following chapters will demonstrate how different groups of experts became invested in thinking about and developing ways of improving girls' health and wellbeing, sometimes supporting medical viewpoints and sometimes challenging them.

2
Reinventing the Victorian Girl: Health Advice for Girls in the Late Nineteenth and Early Twentieth Centuries

By the 1880s and 1890s, even as doom-laden commentary about the trials and tribulations of female adolescence continued to flow from the pens of medical authors, a new model of girls' health began to assert itself in medical and cultural consciousness, reflected in the burgeoning genre of advice literature intended for young women. This new literature on health offered information directly to young women and addressed the question of how girls should approach the expanding opportunities opening up to them in education, the workplace and a variety of cultural, recreational and social arenas, as well as the challenges of adolescence itself. Importantly, it focused less on girls' biological limitations, highlighting the role of their attitudes and behaviour, and the ways in which a positive state of mind and commitment could support their efforts to strive for improved health. Girls themselves were ascribed a great degree of responsibility in their pursuit of good health and it was also suggested that inappropriate behaviour and neglect of the rules of health put their physical and mental health at risk.

Representing girlhood as 'modern' and 'dynamic', commentary on girls' approaches to health and wellbeing was confident and forthright enough to inspire comparisons with the discredited ideals of femininity and the female form of earlier decades. New models of vigorous, bright-eyed girls were depicted, very different from their wan, drooping, nervous and sofa-bound Victorian predecessors, who were described in negative – even mocking terms – for their physical slackness and lack of zest, as objects of so much medical concern and the subjects of so many rest cures. The mid-Victorian girl was depicted as a poor type, a failed example for the modern girl. In 1894 the *Girl's Own Paper* (*GOP*), for instance, contrasted the ideal young woman of the day with her 1850s prototype:

> The young lady with sloping shoulders, gazelle-like eyes, and unchanging amiability would find no place in the present world of women. A course of gymnastics would be ordered as an antidote to her tendency to faint at critical and uncritical moments ... should she return to the 'Book of Beauty', she would find her place usurped by a type, distinct; with characteristics utterly unlike her own. In place of her rounded, irresolute chin, she would find a chin, firm and resolved ... her expression, inane and colourless, would be overshadowed by one of intelligence and character ... where there was only weakness there is now strength and purpose.[1]

The popular periodical *Health* demonstrated how a problem-solving approach to ailing girlhood had evolved. The piece explained how early in the century 'girls were taught that they had no muscles to develope [*sic*], no lungs to expand, nor circulations to stimulate ... Ah me! The imperfect lungs, weak spines, indifferent digestions, bad circulations, hysteria, and a hundred minor maladies that women suffer from are traceable to the old-fashioned physical – or lack thereof – training that girls received.' By 1895 things were very different, the article continued, and young women were taught 'that to be healthy and to rejoice in health and exercise takes no whit from the beauty of girlhood or the womanliness of women; but, on the contrary, they become bright and cheerful, when otherwise they would be ailing and querulous'.[2] The feminist magazine, *The Woman's Signal*, published extensively on women's employment as well as temperance reform, and encouraged women to take up exercise, remarking that ideals had changed about their constitution and habits: 'The notion that a loveable woman must be able only to lie upon a sofa and speak in feeble tones of her fragile and delicate state of health is completely exploded.' 'Health' was now 'a condition of beauty'.

> Hence, every year sees an increase in the number of our girls who play at some active game – tennis, rounders, golf, and even cricket – who cycle, swim and skate. Already the consequences of the improved state of feeling that this indicates are visible. The proportion of girls who are tall and well-formed at present is extra-ordinary. With this development of physique, disease has not disappeared, but 'hysterics,' fainting fits, sudden weepings, and the general *malaise*

and delicacy that used to be so common have all 'gone out of fashion.'[3]

The turn of the twentieth century was marked by the engagement of a variety of 'experts' with the health of young women. Doctors, including a growing band of female practitioners, contributed a large amount of copy on the subject of girls' health to health manuals and periodicals, lay journals and magazines, and so too did social and welfare workers, educationists, youth and club workers, headmistresses, teachers of games and domestic science, social commentators and journalists. Girls themselves were envisaged as a new audience for advice and information. For some, writing for lay outlets allowed for a creative recycling of materials and was a valuable income stream; many advice books sold well, passing though numerous editions, while an expanding range of periodicals offered the opportunity for the regular publication of shorter pieces. To a certain extent, girls' health was contested by these experts, all vying to establish authority to manage the process of adolescence, adding to the ambivalence that already existed amongst medical professionals about the health status of girls and the limits to their activities noted in the previous chapter.

Many of these interactions took place outside of medical arenas, notably in schools, clubs and workplaces, which is indicative of medicine's extended reach into public and semi-public arenas that had previously been largely independent of this form of intervention.[4] It was spurred on no doubt by the broader interest of Victorians and Edwardians in health regimens and sport, and the shifting emphasis onto girls' health was in many ways a natural expansion of the preoccupation with the health pursuits of men and boys that had dominated the mid-nineteenth century.[5] The production of a specialist literature for girls can be seen too as an extension of the long-standing interest of families in implementing health interventions in the home as well the stepped-up consumption of domestic health products.[6] Health advice to girls added to what was already a vast array of domestic medical guides as well as manuals targeted more specifically at wives and mothers.[7] This was part and parcel of 'the burgeoning of a popular literature of family advice both in books and periodicals' focusing largely on children that Carol and Peter Stearns have dated to the 1830s and 1840s, 'as increasing numbers of middle-class families sought advice books or had such materials thrust upon them in the popular magazines'.[8]

Though this new advice literature for girls engaged with the theme of adolescence as a time of risk and challenge as well as opportunity, much of it embodied an upbeat approach, tending to question, rebuff or side-line dominant medical theories that represented girls as physically and mentally weak and incapable, and to offer instead practical, manage-able approaches to boosting health. Maud Curwen and Ethel Herbert, respectively a lecturer in hygiene to the Staffordshire Education Com-mittee and a former gymnastics teacher, set out, for example, to create health rules and exercises to tackle poor physical health and tiredness in their *Simple Health Rules and Health Exercises for Busy Women and Girls*, published in 1912. Their book aimed to see 'the dull eye brighten, the pasty complexion become clear and fresh, the step springy, the expres-sion happy, the appetite good, the carriage of the body upright, and the whole person, as it were, born again'.[9] Curwen and Herbert offered a rounded approach to health that marked out guides intended for audi-ences of girls, promoting the Swedish system of gymnastics and relaying straightforward information on diet, hygiene, dress, overstrain, sleep-lessness and nerves. Their advice, rather than targeting well-to-do young women, catered for girls who needed to earn a living, and were thus compelled in many cases to balance the 'double burden' of domestic duties with work outside the home. Dr Florence Richards' *Hygiene for Girls*, published one year later, devoted a substantial section to germ theory and hygienic practices, but its overall ambition was to urge 'the necessity of developing habits that will bring out the best possibili-ties of the individual both physically and mentally' and to 'give the young the instruction most needed to enable them to build up nor-mal, healthy lives'.[10] Published initially in the US, Richards' book was reprinted in Britain and did double duty as a school textbook and as an advice book for a wider readership. Judging by its lavish illustra-tions, presenting hygienic and poised lifestyles in comfortable homes, it catered predominantly for a middle-class audience of girls. It also strongly emphasised not just individual responsibility for maintaining health, but also the wider benefits to the community of good health practices.

Not all writers reacted positively to the modernisation of girlhood which entailed a more vigorous approach to health and hygiene. Many questioned its potentially negative impact on girls' characters as well as on their long-term health and their role as future mothers, reflect-ing Eliza Linton's assessment of the vulgar and rough 'Girl of the Period', described in the introduction, whose caring character had been degraded by ambition and the desire to pursue a range of activ-ities which took her away from the home and her domestic and filial duties.[11] Mary Whitley encapsulated this concern about the modern girl in her *Every Girl's Book of Sport, Occupation and Pastime*, published in 1897:

> A great change has come over the lives of English girls...No one can deny that the modern girl with her outdoor atmosphere, and her frank devotion to sports and games, is a decided improvement upon the limp and lackadaisical young female of the early days of the present century, who fainted and screamed at the sight of a mouse, and who looked with perfect terror upon a three-mile walk...At the same time, although I am an eager advocate for every kind of physical exercise for girls, I would have them carefully guard and cultivate the gentleness and 'sweet reasonableness,' without which they cannot hope to possess a truly lovable or womanly character.[12]

The 'New Girl', some commentators declared, had lost her self-sacrificing character and feminine virtues, becoming hard, brusque and selfish. Periodical literature presented a variety of views on this theme, even within the pages of the same journals. The hot topic of female cycling had its supporters and detractors, as did the subject of education for women. The anti-modernising and anti-New Woman writer Jennie Chandler, a frequent contributor to *Good Health*, warned in 1894 that poor health and unhappiness would result from girls wanting to be more like men; after a half century of pursuing expanded opportunities for education, it was apparent that 'Nature has put so many obstacles in their way that they can never succeed to any extent, but the effort does harm to health and character.'[13] However, an article published in the same journal one year later was positive about the 'New Woman', who would not, it argued, abandon her natural vocation of bearing and rearing children, but also had great potential in other ways:

> Physiologically, the new woman is a vast improvement over other women. She is not a delicate, pallid, emotional, unknown quantity. She is a creature full of energy and decision. She takes systematic stock of her vital capital, and endeavours in every way to increase it. She lives in a hygienic regular fashion. Eats three hearty meals a day, is fond of athletics, and keeps in touch with the world mentally and physically.[14]

In 1899 the editor of the *Girl's Realm* reflected back on the 'Girl of the Period', as framed by Linton, arguing that she had 'created a monster on purpose to slay it', and talked of how the 'modern girl', the current 'Girl of the Period', was 'a creature of the open air, she wants to be stirring', who adored sports and physical culture, who claimed the right to an education equal to her brothers' and to a career. At the same time, her approach 'must be shaped by reason and commonsense', not by having a 'good time'. 'Take her all in all, the girl of the period is a "fine fellow," breezy, plucky, quick to enjoy, and ready to stand by her sex ... There is forming an *esprit de corps* amongst girls which delights me.'[15]

Exploring health advice to girls from the 1880s through to the early twentieth century, including the particular case study of the *GOP*, this chapter illuminates the increasingly positive health messages conveyed to them across this period, which focused on the cultivation of a confident attitude, the management of body and mind, and attention to beauty, hygiene, diet and appearance. Advice on these topics differed substantially from the medical discourse marked largely by pessimism and anxiety about the vulnerability of young female bodies and their liability to numerous and alarming disorders discussed in the previous chapter, though this continued to shape writing on menstruation, which tended still to be directed towards mothers rather than their daughters, at least up until the 1920s. Advice to girls generally focused on enabling them to achieve better health by means of straightforward instruction. In some instances it tackled head-on the ghastly scenarios set out in the medical literature for ambitious young women who exposed themselves to great risk by taking on new activities, notably higher education. But more often it offered neutered interpretations of alarmist medical models, sidetracked them or provided simple tips on how to reduce their impact.

Advice literature for girls

Carol Dyhouse has suggested that the period 1890–1920 was notable for its heavy concentration on boys and references to adolescent girls were 'value-laden and stereotyped'.[16] However, it could be argued that this period was notable for the separation out of girls as a discrete readership, an aspect of the New Journalism of the 1880s and 1890s, 'with its ever more diversified target groups'.[17] Rather than stereotyping girls, many writers perceived them as dynamic and constantly evolving as they took up new roles. Advice on health management devoted exclusively to girls proliferated, and engaged too with a concept of health based not merely on concern with bodily issues and combating or avoiding disease and illness; rather, health was described as enabling and energising, embracing physical, spiritual and mental wellbeing. And while healthful activities involving girls took place increasingly in public or semi-public spheres, advice literature also strongly promoted perseverance with the rules of health in the home.[18] Ina Zweiniger-Bargielowska has suggested that approaches combining physical and mental health were unusual in the early twentieth century 'and differed from the bulk of the literature which emphasized a desire for beauty as the prime incentive for female physical culture'.[19] This viewpoint will be challenged here. It will be suggested that it is precisely this association of physical and mental health, as well as an emphasis on the ways in which behaviour could shape wellbeing, that distinguish this literature. Besides this, physical culture, rather than being tied to the gymnasium and a few prominent publicists and entrepreneurs, was a label applied loosely and broadly, embracing a variety of approaches to health, exercise and sport, an issue to be explored in more depth in the following chapter.

Concern with the health of young women fitted into a broader process of defining girls as a specific readership group with particular needs and goals.[20] The creation of girls' periodicals was largely an innovation of the late Victorian period, dovetailing with the rise of the 'New Woman' and accompanied by a growth in New Woman literature.[21] This period was marked too by the production of instructional articles, material on careers and human interest stories, intended to urge readers to engage with new ideas and activities, as well as a general feminisation of the press and the appearance of visually more exciting magazines. This era was also associated with the emergence of the ambitious and energetic 'Girton Girl' and a 'more general reworking of the concept of "girl", a process which the periodical press of the 1880s and 1890s both mediated and helped to create'.[22] It saw the labelling of various forms of 'girlhood' or girl types, including the mill-girl or factory girl, girls participating in the newly established girls' clubs, girls interested in the more traditional issues of fashion and romance, and modern sporting girls. The new periodicals – notably the *GOP* (launched in 1880) but also *Young Woman* (1892), *Young Gentlewoman* (1892), *Girl's Realm* (1898), *Girls' Friend* (1899) and various papers aimed at mill girls – capitalised on the new 'girl culture', as during the 1880s and 1890s the girl 'became a contested signifier, creating a problem not only of definition, but, as many writers of the period would suggest, of identity'.[23] The term 'girl' additionally became an aspirational label that would be adopted and promoted by magazine publishers eager to market their new products to an older generation of women, who might wish to prolong their own girlhood or participate in the dynamic activities associated with young women of the period, while some publications directly targeted both girls and their mothers. The proliferation of girls' periodicals was accompanied by a plethora of annuals, books of sport, pastimes and hobbies, and general advice books for girls. Emphasis was placed on young women as a discrete audience, young women whose ties with domesticity and maternity were being loosened, as they engaged with new tasks and functions – work, study, recreation and sport – topics which in themselves delivered up a great deal of interesting copy for editors and publishers.

An increase in periodical publications went alongside rising newspaper circulations, with newspapers containing an array of adverts for health products and articles on a range of health issues. At the same time, newspapers 'discovered' the woman reader and the 'illustrated press put the female body under scrutiny as never before'.[24] Fashion was a major feature of women's pages alongside articles on beauty and beauty aids. Lord Northcliffe, founder of the *Daily Mail* and the *Daily Mirror*, led the way in this process, launching two women's magazines, *Forget-Me-Not* and *Home Chat* in 1891 and 1895, respectively, and set aside columns for women in his newspapers, with features on dress and sewing, hygiene and beauty, cookery and home matters.[25] Magazine advertisements, meanwhile, promised 'transformation of the female body', with pills for reducing weight, or developing larger busts, or to banish nerves and pallor, with patterns of consumption being first formed at adolescence, as this coincided, according to Thomas Richards, with 'physiological changes that issue from the consumer's body'.[26] Adverts focused, for example, on skin care, physical strength and hygiene products, while others, such as one for Carter's Little Liver Pills, alerted the viewer not just to what the pills would cure or the defects that they would remedy, but to the fact that they would create 'perfect health' (see Figure 2.1).

Built strongly into the sales pitch – and increasingly into images used in advertising – were representations of the modern, fashionable, healthy, fit and dynamic girl. Girls became emblematic of the commodification of health, with fresh-faced energetic young women advertising products as varied as soap, cocoa, Beecham's Pills, creams and ointments, and corsets of various designs.[27] Many advertisements promoted products intended to achieve the ideals of healthy girlhood – blood pills for pale-faced girls or loose flexible clothing – or took advantage of the new emphasis on outdoor physical fitness, girls' participation in sports such as tennis or golf, or drew on images of the 'seaside girl' strolling along the seashore, swimming or dozing in the sun. With their exposed limbs and particular poses, many of these were at one and

Figure 2.1 Carter's Little Liver Pills for headaches, biliousness, torpid liver, constipation and indigestion: Display card, c.1910 (Wellcome Library, London).

the same time healthful and 'sexualised'.[28] Other advertisements were deeply at odds with new models of girlhood, notably what appeared to be – despite the protestation of texts declaring 'perfect freedom of action' – highly restrictive corsets that perpetuated artificial images of young women. Many of the adverts were 'spectacular' in their claims as well as their visual imagery, a sharp contrast with the 'normalisation' of health and body image for girls which was emphasised in much prescriptive literature.

By the 1880s and 1890s healthy girlhood had become a regular topic in health periodicals aimed at a general readership, such as *Good Health*, *Health* and the *Domestic Magazine*. In 1894, for example, *Good Health* contained features on reform clothing, hygiene, diet and exercise for young women, callisthenics for girls, and cycling for women. *Health: A Weekly Journal of Sanitary Science*, founded by Dr Andrew Wilson in 1883 to promote ideas of public health, at the same time introduced the notion of individual responsibility for health based on improved knowledge; it included a series on 'Physical Exercise for Women' and articles on athletics for girls, hysteria and the diseases of women. Its column 'Talks with Girls' covered posture, dress and corsets, education for women, swimming and other sports, and healthy habits.[29] Many women's magazines included features on the health of young women, exposing them to yet more information on this subject, for, while a discrete literature emerged for girls, categories of reader were still expected to overlap. The *Queen*, an illustrated and eclectic ladies' newspaper, which kept women up-to-date on politics, higher education and female employment as well as fashion and fashionable society, featured articles on cycling for ladies, reform clothing, girls' clubs and gymnastics in its 1884 issues, alongside adverts for high fashion garments and various forms of corsetry.[30] While the focus of much of this output remained on middle-class audiences, working girls were also targeted by magazines and agencies concerned with their health and welfare, such as the gazettes published by the Young Women's Christian Association, which highlighted numerous health issues in published lectures and other features on hygiene, exercise and nutrition.

Authors and publishers alike appear to have recognised a golden opportunity to open up a new market for information and advice on health targeted at girls as part of a broader expansion in health advice literature during this period.[31] Several authors published whole series of manuals aimed at varied audiences as well as general household guides. Gordon Stables, health columnist to the *GOP* and the *Boy's Own Paper*, published health guides for girls and boys, adults of both sexes, and mothers of infants and young children, as well as books on cycling and health, the health benefits of tea and bathing, and even canine medicine, and was also a well-known author of children's adventure stories.[32] His guide for girls was the first of his books on health, and apparently alerted him to the broader potential of the market. Popular health guides and periodicals pointed their readerships to further publications on health and included advertisements for health products. Gordon Stable's *Girl's Own Book of Health and Beauty*, for instance, promoted his many other publications on health, an advert for lavender water and Professor Dowd's Health Exerciser, a complete family home gymnasium, his *People's A B C Guide to Health* advertised the 'Gordon' Medicine Chest recommended for travellers, cyclists and tourists, Pears' Soap, antibilious pills, gripe water and bicycles.[33]

In many ways, guides on girls' and women's health constituted a refinement, a new category in the production of domestic manuals intended for lay consumption, a production that appears to have been thriving in the late nineteenth century, indicated by the continued appearance of new texts and editions. General domestic guides and encyclopaedias of medicine intended for home use also contained a good deal of information on girls' health, focusing particularly on the management of puberty, and broaching new themes and addressing current debates, such as the impact of secondary education on young women. George Black's *Household Medicine*, first published in 1883, confined itself to a discussion of the appearance of menstruation and green sickness or chlorosis, thus framing the health of adolescent girls in terms of anxieties about their reproductive bodies and the challenges posed by puberty.[34] However, other volumes, such as Dr Malcolm Morris's multi-authored and encyclopaedic *Book of Health*, published in the same year, embodied a more dynamic approach. Intended for the 'general reader', it included sections on growth, periodicity, clothing, physical education and exercise, recommending active muscular work for young women and hygienic dress, and contained a great deal of commentary on the consequence of study for girls (representing both sides of the debate) and the health of working girls.[35] The Olsens' *School of Health*, published in London in 1906, was inspired by hydropathist and food reformer John Harvey Kellogg's robust approach to health (the Olsen's also edited *Good Health* magazine). Their book stressed the importance of the principles of hygiene and natural approaches to sickness, including hydrotherapy, and included tips on beauty, dress and cleanliness for young women, a huge amount of information on food and diet, as well as advice on the management of neurasthenia and hysteria.[36]

A number of books specifically addressing the health of young women made their way across the Atlantic, bolstering the steady output of British publishers. Volumes written by two American physicians: Anna M. Galbraith's *Hygiene and Physical Culture for Women* and Florence Harvey Richards' *Hygiene for Girls*, were both, for example, reprinted in London.[37] This formed part of a much larger and durable two-way transmission of health advice across the Atlantic, which also included a good deal of information on new systems of health such as hydropathy as well as the physical culture movement.[38] Other volumes were produced by global publishing networks, such as Mary Humphreys' *Personal Hygiene for Girls*, published by Cassell in 1913 in England, North America and Australia.[39] Health periodicals relied on overseas correspondents to write many of their columns, while debates, for example, on the phenomenon of the New Woman or women and higher education, crossed national borders, exposing readers not just to new ideas on health, but also giving them access to a range of issues connected with modern girlhood in a wide variety of contexts.

Notably, many of these authors were women, as a steadily growing corps of women doctors fed into the production of advice literature. The emergence of healthy girlhood coincided with a rise in the number of women qualifying in medicine, many of whom went into careers dedicated to the health of women or children – obstetrics and gynaecology, paediatrics, public health, school medicine, domestic science, factory inspection and health education.[40] Though Brian Harrison has questioned the impact of female practitioners on women's health given that there were so few of them, they appear to have made an impression out of all proportion to their numbers. Aside from treating women and children in their practices, they contributed numerous publications – periodical articles, pamphlets and books – on the health of women and girls, and many were active on the lecture series circuit.[41] After the turn of the century a cluster of medically qualified 'feminist eugenicists', including Arabella Kenealy, Mary Scharlieb and Elizabeth Sloan Chesser, produced an array of advice literature for women and girls and lectured to a wide variety of audiences. Chesser, for example, wrote for schoolgirls and women at different life stages, offering information on health, motherhood and the psychology of sex in practical psychology journals, newspapers, women's magazines and advice books.[42] While preoccupied with the role of girls as future mothers in securing Britain's racial integrity and the health of the nation, Chesser also advised them to engage broadly in civic and social life, and provided instructions on how to harness health to produce greater vitality.

Mothers, daughters and the epoch of puberty

Increasingly, girls' experiences were regarded as having less in common with those of their mothers – and, after all, the sofa-bound mid-Victorian girls who were depicted so negatively in much of the literature *were* late Victorian girls' mothers. Prescriptive literature began to speak directly to young women, who were held responsible for their own health, a process underlined by the images contained in books and periodicals of brisk young women going about their business, dressed in modern attire.[43] Advice books speaking first and foremost to mothers and adopting the rubric of so much Victorian self-help literature – 'an invaluable work for mothers and daughters' – started to be substituted by guides catering directly for girls. The idea that mothers should act as conduits of advice, however, endured in some manuals, such as Pye Henry Chavasse's *Advice to a Mother on the Management of Her Children*, which was addressed primarily to mothers though it also spoke to the 'rising generation'. It took the form of questions and answers to the reader and Chavasse stressed that the book 'ought not to be listlessly read, merely as a novel or as any other piece of fiction, but it must be thoughtfully and carefully studied, until its contents – in all their bearings – be completely mastered and understood'.[44]

An important exception to this increased emphasis on informing girls directly about their health and physiques was the management of menstruation, and doctors continued to highlight the role of mothers in preparing girls for its onset and in dealing with menstrual hygiene, topics broached only occasionally in health guides for girls before the 1920s and 1930s.[45] In 1852, obstetrician and gynaecologist Edward John Tilt had found that the vast majority of girls were unprepared for their first period, and chided mothers for their neglect, and doctors continued to lament the failure of mothers to provide adequate instruction and support for this momentous phase.[46] Menstrual hygiene was to be managed in the private sphere and mothers were to have 'power over the arrival of this crisis'.[47]

The more traditional format of advice books intended for young wives, such as Lionel Weatherly's *The Young Wife's Own Book* and George Black's *The Young Wife's Advice Book*, as well as general domestic advice manuals, typically contained a brief summary of the management and hygiene of menstruation and some information on its health risks; a far smaller number also alluded to 'pollution' or 'menstrual etiquette enforced by ridicule and ostracism'.[48] In his *Domestic Medicine and Surgery*, Dr J.H. Walsh urged mothers to inform their daughters about the changes wrought by puberty, otherwise their daughters might feel shame about menstruation and attempt to repress it. Walsh suggested that menstruation was normally healthy, though 'anything which interferes with the general health is found to affect this secretion'. He additionally outlined the chief symptoms and treatment of chlorosis and anaemia in young women, who he described as 'unimpregnated females', recommending iron, pure air and exercise as essentials of treatment.[49] George Black commented on girls' sudden growth spurt at puberty, causing 'more rapid developmental changes, and greater strain on the constitution in girls than in boys. Hence the great importance of adequate nourishment at this period of life to prevent anaemia.'[50]

A number of physicians recommended the curtailment of physical exercise, cautioned against travelling, urged avoidance of over-exciting entertainments and activities at puberty, and dwelt too on the physical, and more implicitly, sexual, changes taking place at puberty. Lionel Weatherly's *The Young Wife's Own Book*, published in 1882 as one of a series of popular handbooks on household medicine and health, warned of the danger of following music lessons during menstruation; sitting upright with the back unsupported and 'drumming vigorously' on the piano was detrimental, resulting in 'strong excitement of the emotions'.[51] 'The borderline between childhood and womanhood', was described as 'a most trying time, both as regards mental and bodily health'.[52] Weatherly advised against taking violent exercise and urged girls to refrain from cold baths and sea bathing. His book provided information on married women's health and that of her offspring, its advice extending across the female life cycle, covering puberty, menstruation, pregnancy, lying-in and the management of health during the change of life. It was apparently intended as a durable text, which, while focusing chiefly on young women in transition from girlhood via wifehood to motherhood, would guide mothers and their daughters through their reproductive careers. Weatherly also expanded on the various disorders of menstruation, which in his view were a frequent cause of hysteria: 'The number of colourless, pale-complexioned, short-breathed, and sickly-looking girls that are ever to be seen, own their condition, in nineteen cases out of twenty, to some unhealthy form of menstruation, which if not taken in time may soon land the poor creatures into a consumption.'[53] Aside from the management of menstruation, Weatherly cautioned mothers to monitor their daughter's interest in the opposite sex with great caution. Ideally, puberty mapped the change 'from devoted daughter to perfect wife and mother', but it could be derailed into 'an absorbing passion, which entails loss of name, fame, family, and self respect'.[54]

George Black's *The Young Wife's Advice Book*, described the passage from girlhood to womanhood and the commencement of menstruation as being marked initially by languor and dull pains, a feeling of dragging and weight around the small of the back, and dark rings under the eyes, which passed away as menstruation became established, as the girl's frame filled out and her demeanour altered; 'Her bearing becomes more dignified; she exchanges the pursuits of girlhood... she becomes more retiring. It seems as if a great mental change had come over the girl, and there had begun to dawn upon her mind the consciousness of that important mission she is destined to fulfil.'[55] Girls of the upper classes, in Black's view, could anticipate the earlier onset of menstruation compared with lower-class girls, who took more muscular exercise, ate plainer food, had less mental excitement and whose surroundings were 'more conducive to the development of a healthier and hardier frame'.[56] There was considerable ambiguity on this theme; poor nutrition tends to delay menarche, and thus it was likely to commence later for working-class girls, yet for Black and other writers on this subject, late onset of menstruation was indicative of a state of health rather than disease. *Wood's Household Practice of Medicine Hygiene and Surgery*, published in 1881, explained how at the onset of puberty, 'the pelvis enlarges, the hips are expanded... the breasts become fuller... the entire figure loses the angular and awkward outlines of girlhood and becomes graceful and beautiful... new duties and responsibilities devolve upon her'.[57] It was asserted that normal puberty should cause little concern and required only rest, a simple diet and warm hip-baths. In the case of premature puberty, moderate outdoor exercise was advocated, nutritious food and tonics, cold baths and avoidance of anything 'calculated to arouse the passions, such as going into gay company, reading exciting works of fiction, etc.' 'The artificial maturity induced by these and similar agencies is always to be deprecated, for the result is very likely to be a lasting injury to the system.' In the case of the non-appearance of menstruation, Wood's guide urged a rich diet, a great deal of open-air activity, preferably in the countryside or at the seaside, and gymnastics appropriate to the girl's strength. Mothers were urged not to dose their daughters with domestic remedies to bring on menstruation, as this was dependent not just on age but also the maturity of the organs.[58]

Generally, the literature skirted around the physiology of menstruation, itself poorly understood at this time, and the topic of blood loss was rarely broached. Gordon Stables gave a guarded summary of menstruation in his health guide for girls though he chastised mothers for keeping daughters in the dark and urged them to pay particular attention at this point to their health and cleanliness. He tackled the 'delicate subject' with unusual frankness, however, in his *Wife's Guide to Health and Happiness*, describing the duration and volume of bleeding, and stressing the importance of a frank and open mother–daughter relationship:

> But there are many cases in which this confidence does not exist, and constitutions have been ruined for the want of it. A child begins to menstruate, and does not know what is the matter with her, and though alarmed at the sight of blood, she fears to speak of it, or confide in her whose motherly help and advice are wanted now more than ever. Often these three words: '*It is natural*,' would raise a girl to joy from the very depths of despair.[59]

Stables described menstruation as 'the opening of one of Nature's safety valves'. Normally nature could be left to take care of itself, although girls were cautioned to avoid vigorous exercise and excitement, to keep themselves warm and omit their cold morning tub. Stables went on, 'a medical man can usually tell a really healthy girl at a glance, or one, who when married, is likely to be the mother of a bright and happy family'. She should be cheerful and healthy, broad shouldered, 'bust just pronounced enough, but not too much so', wide hipped: 'This is what a physician would call a really promising girl, and he would be ready to give his word for it, that her monthly periods occurred without a hitch.'[60]

When information on sexuality and menstruation was passed on by mothers, it could trigger distaste or despair. Helen Corke, born in Hastings in 1882 to a modest shopkeeping family, experienced revulsion when her periods started, for which she had no preparation. At that point her mother proceeded to give directions on personal hygiene and 'hastily, to advise me of all the precautions and prohibitions relative to the monthly period that she had herself received at my age...For the first time I wish, heartily, that I were a boy.'[61] She went on to discover a copy of *Esoteric Anthropology* in her mother's linen drawer, which explained the 'whole sequence of coition, conception, generation and birth. Wonderfully fascinating – at the same time alluring and repellent.'[62] Edward Bruce Kirk's *A Talk with Girls about Themselves*, published in 1895, was unusual in offering frank advice on sexuality and intercourse (including a glaring anatomical error), even if it was couched in a more lyrical language than his advice to boys. Two options were offered to parents in explaining the 'mysteries of life' to girls, and the book came with perforated pages that could be torn out in making this selection. Kirk advocated the more explicit option as, in his view, girls had the right to know the facts of life and to understand their bodies:

> At marriage, when the man and woman closely unite in loving embrace, the life fluid passes from the man's sexual organ into that of the woman, the entrance to which is the passage in front of the body. This is used, as you know, for the passing out of urine, but it also leads to the womb, which is the magic chamber into which the life-germ ascends. Here the male cell unites with the female seed cell and fertilizes it[63]

Well into the twentieth century, technical guides intended for instruction in girls' schools and colleges, while clearly outlining every other form of bodily function, including respiration, circulation and digestion, said little on the topic of menstruation or reproduction more generally. Elizabeth Sloan Chesser's 1914 volume intended for girls of secondary school age, referred merely to the need to take special care to guard against chills and to avoid excessive physical fatigue:

> Whilst most girls do not experience any adverse symptoms during this period, others may feel less fit, physically and mentally, and it is certainly advisable in these circumstances to avoid excessive mental strain, and to rest as much as possible. Cold baths ought not to be taken... At the same time, there is no reason for any girl of normal health to alter the day's routine or to give up her usual work or recreation.[64]

A great deal of advice literature, including that directed at girls themselves, turned, by contrast, to the themes of chlorosis and anaemia, disorders associated with the poor management of menstruation. *Good Health* referred to chlorosis as a disease of early womanhood, which usually struck girls aged between 16 and 20; it was serious, but curable with doses of iron, bed rest, massage, good diet and, where possible, a seaside holiday.[65] Chavasse directly attributed chlorosis and the suspension of menstruation to poor management and, particularly, a poor diet: 'The disease generally takes its rise from mismanagement – from Nature's laws having been set at defiance.'[66] Anaemia was a 'trying complaint' according to Curwen and Herbert's *Simple Health Rules*, lowering 'the vitality so that life becomes a burden'. Fresh air and exercise were recommended and as well as iron jelloids, though these, they cautioned, were likely to cause constipation and it was vital that the bowels be kept regular, another preoccupation of advice literature.[67] Margaret Hallam pitied anaemic girls who were unable to exercise without fatigue, with their complexions of 'waxy whiteness'. 'Fresh air, food and rest are three of the chief weapons used to fight the foe, assisted by a judicious course of tonics; and, as the remedies are simple, the cure lies greatly in each girl's own hands, unless the symptoms have become very far advanced indeed.'[68] Dr Emma Walker's *Beauty Through Hygiene*, published in 1905, was one of the few guides directed at girls which discussed menstruation or what she referred to as 'the periodic illness'. Walker took a brisk approach to its associated symptoms, referring to lassitude, general discomfort, headaches and dark eyes, normally nothing alarming, though she advised girls not to make serious decisions during the 'menstrual week'. Moderate exercise was approved of, but not excessive muscular exertion. Walker also discussed 'menstrual decorum', urging self-control when feeling irritable as well as keeping 'the parts most concerned... carefully washed... to keep yourself sweet and dainty'.[69]

'Lay down rules and stick to them': girls, advice and responsibility

Even though it largely denied them direct access to information on one of the most important aspects of their health and bodily functions: menstruation, advice literature catering specifically for girls aimed to imbue them – whatever their social class – with a sense of responsibility for their own health. Advice manuals, whether intended to offer advice to 'young wives' or the new genre of material catering specifically for girls, was prolific and enduringly popular. Marianne Farningham's *Girlhood*, first published in 1869, was widely circulated and specifically advertised as suitable for Sunday School prizes, and Farningham also pointed out that her book passed from generation to generation within families. By the time a revised edition appeared in 1895, with a greatly expanded section on health, some 25,000 copies had been printed.[70] Advice to girls addressed numerous and various health issues – exercise, sport and physical culture, diet and nutrition, personal, domestic and public hygiene, dress and beauty, and mental and spiritual wellbeing. Girls were alerted specifically to the dangers of tuberculosis, which was depicted as resulting from a lack of attention to the rules of health. They were also encouraged to develop skills in home nursing and the treatment of minor injuries. Strikingly, girls were rarely enjoined to seek medical support in their pursuit of health; a book, a regime and a determined attitude would be sufficient. Indeed, Dr Gordon Stables criticised girls who suffered from 'bloodlessness' or anaemia for hanging on to the skirts of the doctor's toga, and being 'but little inclined to do much for themselves, although their restoration to health rests so much on their own end of the lever'.[71]

Deborah Gorham has argued that advice books for Victorian girls urged them to become more restrained as they approached adolescence, yet also impressed upon them the need to care for their own health.[72] By the late nineteenth century the emphasis on girls being responsible for their own actions and their own health was stepped up, and the ideal of 'restraint' substituted by the notion that health was potentially enabling, and, for some girls, necessary to earn a living. Gordon Stables referred to the health of working girls as a form of capital to be built upon, while Curwen and Herbert's guide for busy women and girls was at pains to tackle the tiredness that overwhelmed the milliner, the seamstress, the clerk and the shop girl.[73] Health advice to girls not only focused on a wide range of subjects, but also saw them as linked; only girls who tackled all their deficits, and took care to improve their approach to diet, exercise, rest and hygiene would achieve good health. Healthy habits involved brisk daily regimes – cold baths, appropriate attire, changing wet boots, avoiding snacks between meals, creating time for rest and recreation, leaving windows open at night, and being meticulous in the preparation of food.[74] The guides urged common sense and patience. Girls were advised to start steadily with new exercise regimes, not to expect overnight results and to strive to overcome weaknesses, minor complaints and self doubt in their pursuit of health.

Gordon Stables insisted in his *Girl's Own Book of Health and Beauty* that girls themselves take responsibility for implementing his directives. His years of experience in producing an advice column for the *GOP* made him, in his view, the best judge of 'what young ladies want to learn, and I have based the instruction given in this book on such knowledge'.[75] His book embraced the themes of diet, beauty care and exercise, sleep and sleeplessness, and sick nursing, 'things our "Nursie"' ought to know, delivered in a brisk no-nonsense manner. Dr Emma Walker's *Beauty through Hygiene* offered information on a range of sports and exercises, breathing, bathing, care of the hair, eyes, mouth and teeth, clothing, diet and sleep, constipation, the periodic illness, and mental wellbeing or, as she put it, 'cheerfulness', and carefully outlined daily routines of breathing, bathing and exercise. She advocated deep breathing as an aid to health and strength, describing a method of inhaling positive qualities, love, wisdom, usefulness and power for good, while exhaling prejudice, weakness and folly.[76] In advising girls who felt out of sorts, tired and spiritless, Gordon Stables set out a full action plan. He suggested that girls carefully consider how their health could be restored, and also what rules of nature they may have contravened; they should keep a notebook recording their scheme of treatment, attend to their diets, take regular baths and plenty of exercise 'and I insist upon it that it shall be of an exhilarating nature. I will not allow you any long, moping walks all by yourself.' Time should be made for recreation, girls were to ensure that they slept in well-aired rooms, and avoided medicines. 'Lay down rules and stick to them, and I believe in a month you will be able to say, that it was a lucky day for you when you procured this book.'[77]

While Stables apparently disregarded the notion that girls had a fixed fund of energy with which to work, suggesting that application of his health advice would bolster strength and vitality, Walker emphasised the dangers of overexertion, though for her it was exercise rather than brain work that imperilled girls' wellbeing. Walker's book was prefaced and edited by the feminist eugenicist Arabella Kenealy, who explained that 'Each person is able to produce only a certain daily amount of nervous force...if we allow too much to the muscular system, the brain and other qualities will be deprived.'[78] Fellow eugenicist Elizabeth Sloan Chesser warned of the dangers of over-fatigue, but, taking a rather different tack to those physicians who argued that energy supply was limited, claimed

> We have all got a reserve of energy that can be drawn upon, but it is never wise to *over*-draw our health account. We should rather, by studying health and hygiene, by good management of our hours of work and our hours of leisure, conserve our energies and expend them to the best advantage.[79]

In their *Health Reader for Girls*, the Stenhouses likened the human body to the operation of a steam train. Good food supplied fuel, and exercise, rest and sleep built strength and endurance. In turn, activity provided further mental and physical stimuli, increased appetite and thus spurred the process of energy production once again: 'the blood rushes about with its loads of food and air, keeping **the fire of life** burning at the proper rate in limbs and brain and everywhere else in the body'.[80]

The expectation embedded in much of this information was that girls would participate in education, working and public life, be more visible and more active, and increasingly health advice related to these new contexts. Parents and teachers were urged by Dr Clement Dukes, who wrote extensively on the risks of overstrain for boys and girls, to 'instil into girls' minds the fact that it is their duty to be physically strong, and to provide for its attainment by adequate means', principally exercise. Dukes argued that a more 'reasonable system' of training would banish nerves and fainting in girls.[81] The *'great defect'* of 'the utter neglect of *physical education'* in most schools for girls, he opined, was the very way 'to produce *girls* who are useless as companions, *wives* who live on a sofa, and *mothers* who are unfit for their duties. Besides, if greater pains were taken to make vigorous bodies, we should have more vigorous brains, so that a double gain would result.'[82] This view was confirmed by Elizabeth Sloan Chesser in 1913, who, despite her interest in promoting healthy motherhood, explained:

> A generation ago very few girls thought of asking themselves, at fifteen or sixteen years of age, what they were going to do with their lives. The idea of a girl following any business or profession was undreamed of, but it is becoming more and more usual for young girls to decide upon some special line of work, just as if they were boys.[83]

A positive attitude and the management of nerves

Mathew Thomson has dated the feminisation of psychology and the recognition that women might have distinct psychological concerns to the mid-1930s. However, already by the late nineteenth century a considerable amount of attention was being paid to girls' 'mental hygiene' and psychological wellbeing.[84] The management of nerves, temper, feelings and emotions was a popular topic in advice literature. Nerves, negative emotions or feeling out of sorts was attributed increasingly to physical weakness, neglect of exercise, inadequate rest or inappropriate behaviour, rather than being tied to biological vulnerability or the crises of puberty. Meanwhile, poor physical health was blamed on a poor state of mind and listlessness. Nerves were to be 'managed' and cheerfulness pursued in a similar way to physical health. Stables claimed that nervousness resulted from being run down and impoverishment or poisoning of the blood, caused by overwork, too sedentary a lifestyle, over-indulgence of food or drink, or other excesses. Early rising and sufficient sleep, fresh air, a regular and balanced diet, exercise and in some cases a tonic medicine were, according to Stables, the best remedies for nervous attacks.[85] Alongside information on treating such mundane disorders as pimples, sore throats, nose bleeds, gum boils, warts and smelly feet, as well as how to guard against tuberculosis, Chavasse's *Advice to a Mother* identified torpor and debility of frame, poor diet, and want of air and exercise as precursors of chlorosis and hysteria. Robust diet, fresh air, active exercise and avoidance of 'fashionable' amusements was recommended: 'to employ her mind with botany, croquet, archery, or with any outdoor amusement' and 'above all, not to give way to her feelings, but, if she feels an attack approaching, to rouse herself'.[86] Chavasse's durable volume insisted on paying attention to 'Nature's laws', and claimed that the principle causes of weakness, nerves and unhappiness was 'ignorance of the laws of health, Nature's laws being set at nought by fashion and by folly, by want of fresh air and exercise, by want of occupation, and by want of self-reliance'.[87] For Mary Humphreys, the response to tiredness was to 'draw up your body to a proper posture, open wide a window, and breathe slowly and fully deep breaths of pure air'. Poor posture which had become a 'fixed habit' could be overcome by regular exercise: '**An upright carriage**, head well poised, chest fully expanded, is an outward physical expression of a character fearless in its uprightness and integrity of purpose'[88] (see Figure 2.2).

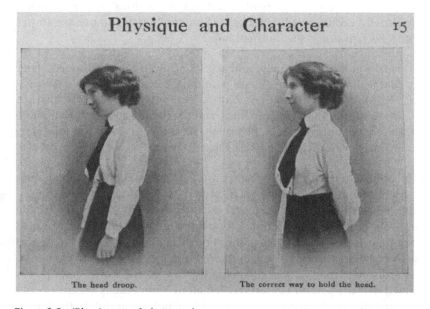

Figure 2.2 'Physique and character'.
Source: From Mary Humphreys, *Personal Hygiene for Girls* (London, New York, Toronto and Melbourne: Cassell, 1913), p. 15 (author's collection).

Elizabeth Sloan Chesser recognised the balance between mental and physical activity, work and meaningful recreation, as 'one of the principles of health':

> It is one's duty to cultivate the capacity for enjoyment, the ability to find happiness in the simple pleasures of life. Cultivation of happiness is a duty, because it not only makes for personal health, but it adds to the welfare and pleasure of the people one comes in contact with ...

> *Mental exercise*, or intellectual work, is another factor in personal hygiene. A girl must do a certain amount of mental work is she is to keep in health. There are very real dangers in idleness. In the first place, a girl who has 'nothing to do' dissipates her energies in the futilities and trivialities of life. If she has not got some real work on hand, she may spend her energy on undesirable occupations, or gradually come to find idleness supportable. This is the beginning of deterioration of character.[89]

Taking a less positive stance, Amy Barnard underscored the risks associated with the mismanagement of feelings or 'passions', the mind being accorded a critical role by Barnard in shaping health and disease. In her *Girl's Encyclopaedia*, she outlined how:

> Passion lowers the vitality and acts like a poison. Unrestrained anger and jealousy upset the digestion, and have been known to drive people insane; so does fright, and it may cause jaundice. Worry disturbs the action of the heart and lung, and also upsets the digestion; while excessive study brings on headache. We see from these examples how intimately mind and body are related.[90]

In comparison with medical texts, the ferocity and impact of hysteria tended to be dampened down in prescriptive literature. Though considered potentially disabling, and likely to severely disrupt family life and challenge girls' aspirations for achieving good health, it was also regarded as treatable, largely without medical intervention. Hysteria was addressed in *Health*, as it was in many other journals and advice books, as a disorder that warranted sympathy, but, at the same time, it was urged that the patient be 'severely left alone'. Sometimes 'a glass of cold water suddenly dashed in the patient's face' helped, as did, in the longer term, fresh air, exercise and healthy pursuits to engage body and mind.[91] Dr A.T. Schofield, a regular contributor on a range of topics pertaining to girls' health, described in *The Leisure Hour* how hysteria's victims were typically under-worked, unhappy, suffering mental strain, exposed to excessive luxury or an ill-balanced education. The condition was often found, he suggested, in people of strong minds with plenty of nerve; the secret of curing all hysteria was 'feeling true sympathy from the conviction of the reality of the disease, but showing none'.[92]

Sleep was accorded huge importance in managing and boosting health. 'Then as the muscle rests, the blood brings more food and oxygen to the tired and hungry muscle cells, and sweeps away their rubbish', so too, explained the Stenhouses, 'When we are asleep, most of the brain, most of the nerves, and most of the muscles are resting. Then they have a good chance of taking up new food from the blood.'[93] Sleeplessness was harmful and meant that the body was unable to repair itself. This was a forerunner of depression, according to Curwen and Herbert, but a change of scenery and attention to the laws of health would put this right. 'Regular exercise, with its improvement of the cir- culation, takes away poisons from the system and improves the outlook of the mind.'[94] Dr Emma Walker suggested that loss of sleep, along with overwork or anaemia and chlorosis, was a fertile source of 'the blues'. She opted to focus on the cultivation of cheerfulness rather than hys- teria or nerves: 'When you smile, mean it', don't scowl, guard your thoughts, 'Walk in the sunshine and its light will be reflected in your faces.'[95] Margaret Hallam encouraged girls not to

> forget that the depression and misery that you suffer comes, not from other people or your own feelings, but simply because the blood is out of gear, and is letting you know it. Say to yourself, 'It is not the world that is awry, it is my own works that are wrong.' Think as little about yourself as possible, and take an interest in other people and other things.[96]

Beauty and hygiene

Even guides that lingered over beauty tips and body care situated this within broader goals of producing healthy bodies and enjoyment of good health, and emphasised the importance of exercise as well as the fostering of good behaviour and character. In 1886, Anna Bonus Kingsford collected together her 'Letters to Ladies', which had been published between 1884 and 1886 in *The Lady's Pictorial*, and repro- duced them in a volume on health and beauty for women.[97] According to Kingsford, her letters had generated a great deal of correspondence which inspired her to remodel them in a manual for popular use. She placed strong emphasis on her medical diploma, which qualified her to write the book, and presented it as the first guide 'to instruct her sex on matters connected with the improvement and preservation of physical grace and good looks'.[98] Its contents were largely dedicated to 'the culture of beauty, grace, and health', hair care, the complexion and the figure, but also included chapters on baby care, and the hygiene and cuisine of the sick room. A large section of the book spoke to 'Dear Sibyl', a mother concerned with the 'care and culture of beauty in her children'; other letters spoke directly to young women concerned about being overweight or underweight and making the best of their appearance. Kingsford, a spiritualist and theosophist, anti-vivisection campaigner, supporter of the women's movement and advocate of vegetarianism, provided much commonsense advice in her letters, which aimed to empower women by improving their health and looks, freeing them from the need to consult with male physicians for what they might regard as 'trivial' matters. Kingsford also widely advertised her beauty products in advice books and the press.[99] *Beauty and Hygiene for Women and Girls*, published anonymously in 1893, included sections on exercise and tips on how to develop a comely figure, dress, and hygiene, as well as a battery of information designed to tackle poor complexions, moles, warts and superfluous hair, and tips on perfume, make-up and hair care, and advertised Dr Anna Kingsford's 'celebrated preparations' to improve complexions and to enhance the bathing experience. It also stressed that 'To possess the charm of loveliness is the legitimate ambition of every women, for physical and mental beauty are inseparable from physical and moral health. No woman, indeed, can be truly beautiful unless she can be healthy.'[100]

The means to achieve beauty were best established in youth, according to Kingsford: 'For the body as well as the mind is most susceptible of impression and training in early years, and . . . the habit of beauty is easier to acquire then than in any subsequent period of life.'[101] 'Under the head of "beauty"', she included, 'of course, health and good sense, for no boy or girl can be really beautiful who is either sickly or foolish'.[102] Gordon Stables concurred that

the girl who was indifferently well is self-conscious, ill-at-ease in society, not clear in eyes, and very often sallow as to skin . . . she cannot smile the smile that wells up from the heart, and goes curling round the eyes . . . No, she cannot smile, she can only make faces with her mouth . . . The wise will, I humbly hope, find much in these pages [his own book] to benefit by, the foolish will go wading through in search of nostrums and recipes to ensure them an artificial beauty.[103]

In 1895 a piece in *Good Health* declared that 'Paints of all kinds spot the skin, and they frightfully distort and destroy the natural beauty of the human face...There is no expression in the face bedaubed with paint.' The article went on to explain that 'A mild diet...good digestion, out-door exercise, and a tranquil mind pertain to loveliness...Perfect beauty cannot exist except there is an interest in something, some occupation or labour.' Diet was crucial, the article continued:

> If a girl sits down to a potato and pickles, strong tea, pies, cakes, ices, and fiery condiments, she will not hold her beauty. As a result, when the girl is twenty her eyes are dull, teeth yellow, gums pale, lips wan, flesh flaccid, and skin unyielding. Recourse is had to padding, face washes, stains and belladonna.[104]

'No amount of "making up"', Amy Barnard declared, 'can replace the glow of health in a clean skin, the gloss of well-nourished hair, and the full development of trained muscles. The girl who would be attractive to look upon must be good throughout.'[105]

Rigorous hygiene was promoted for its benefits to beauty, mental and physical health and as part of girls' daily regimes. Stables advocated the benefits of pure water and encouraged girls to take a sponge bath every morning, and to be particular with their teeth and hair.[106] Even having a good head of hair, according to Stables, was promoted by exercise and a cold bath, as well as 'a calm and unruffled frame of mind'.[107] Curwen and Herbert recommended hot weekly baths and a daily tub or brisk rub down with a damp rough towel,[108] while the Stenhouses devoted two full chapters of their *Health Reader for Girls* to cleanliness and a further chapter to dental care: 'No self-respecting girl can bear the thought of her own body remaining dirty.'[109]

> Amongst the essential principles of health, we must include such hygienic factors as the breathing of pure air, the eating of pure food, the maintenance of cleanliness of body and purity of mind, the obtaining of regular outdoor exercise, and the accomplishment

of definite work, combined with the regulation of rest, recreation and sleep. It is in youth that good habits can most easily be established.[110]

Florence Harvey Richards explained how hygiene of the muscular system was promoted by exercise and that a well-managed body would produce more energy:

The contraction of muscles all over the body during vigorous exercise squeezes the lymph and the blood through the veins to the heart, thus improving the circulation and getting rid of waste matter lying dormant in the tissues. Exercise causes deeper breathing, and the increased supply of oxygen in the lungs purifies more blood, thus providing a larger quantity to be carried to the surface of the body. This induces free perspiration and thus rids the body of more impurities. Exercise uses up food and creates an appetite for more food.[111]

Hygiene, for Richards, referred to the functioning of the whole body, and also required girls to learn about disease and germs (particularly in connection with tuberculosis) as well as sexual hygiene.[112] While tuberculosis was not directly inherited, Mary Humphreys argued, the tendency to acquire it may be, but 'by building up the system into perfect health, and strengthening the lungs by exercise and breathing pure air, much can be done towards resisting the disease and throwing it off in its earlier stages'.[113]

Alongside personal hygiene, many guides provided considerable detail on public health and hygiene within the home, which would ultimately be the responsibility of girls working for the public as well as private good. Housework, as a bonus, was deemed an excellent form of exercise. Margaret Hallam's course of physical culture, published in 1921 and based on the health and beauty articles that she had been publishing for the past 20 years, recommended 'the broom for beauty and the brush for blues', urging that girls, whatever their position in life, be trained from childhood to keep their own rooms clean and dust free. In later life, housework 'was extremely good for the figure, and rounds the arms and takes off superfluous flesh ... with astonishing rapidity', while 'Polishing was one of the finest exercises in the world for developing the arms and chest.' Housework also stimulated the liver and cured the blues.[114] Mary Humphreys insisted on good ventilation and sunlight in the home, as well as 'scrupulous care' in cleaning; 'whatever the status of the home, from the poorest cottage to the palace, there is always some need of the girl, something she may do'.[115] Chesser's *Physiology and Hygiene* was directed primarily at secondary schoolgirls, and provided a great deal of instructional material on food hygiene and preparation, home decoration and furnishing, sanitation and drainage, as well as baby care, sick nursing and the treatment of minor ailments.[116] Germs, or at least 'enemy germs', were to be resisted though 'perfect health', which came down to not only cleanliness and a meticulous approach to housework, but also 'plenty of fresh air, plenty of sleep, and plenty of good, plain, nourishing food'.[117]

Responsibility for the care of infants was an important aspect of girls' duties – de facto in the homes of the poor where they often became substitute mothers – and many health guides included a great deal of instructional material on feeding and caring for babies, with emphasis on this role stepping up in the approach to the First World War as concern with the future generation of recruits became embedded in the campaign to reduce infant mortality. The preface to Annie Burns Smith's *Talks with Girls upon Personal Hygiene*, published in 1912 and directed at girls of elementary school age, alluded to the general health improvements which had taken place in the last half-century, resulting in a decline in deaths and illness. The next step, she argued, was to educate children in healthy practices.[118] Smith, a special science teacher appointed by the Birmingham Education Committee, covered a range of topics in her volume, germs and dust, personal and domestic hygiene, digestion, and care of the teeth, hair and clothing. She concluded with several chapters on baby care and infant feeding, thus emphasising the link between girls' management of their own health and that of the next generation.

Management of the body and its appearance

Physical culture was to have a significant impact in Britain around the turn of the century, promoted by Eugen Sandow, Bernaar Macfadden and a number of entrepreneurial directors of gymnasia that were springing up in major towns and cities across Britain. It became a popular subject in health periodicals and girl's magazines.[119] However, as the following chapter will demonstrate, physical culture constituted only one aspect of the exercise and sporting regimes on offer to women and girls around this period and one of many conduits of advice on health and training. Physical culture shared in the philosophy of promoting an all-embracing approach to health, echoing earlier recommendations on hygienic practices and the cultivation of beauty.

A perambulating wind-mill.

Figure 2.3   Gymnastic dress illustrated in 1893. 'Perambulating Wind Mill'.
Source: From Ellen Le Garde, 'Outdoor Sports for Girls', *Our Own Gazette*, X: 117 (September 1893), p. 106 (Modern Records Centre, University of Warwick, MSS.243/5/4: YWCA/ Platform 51).

As a special British edition of *Woman's Beauty and Health* declared in 1902:

Many people think that 'physical culture' means to don a gymnasium suit and go through a few calesthenics. It means so much more than that. It means the science of living; how to be strong mentally, morally and physically; how to take care of the body; how to *exercise*; how to rest. It is learning to control one's nerves. It is the science of becoming acquainted with one's self; to know how to get the best out of life ... People eat too much, eat too great a variety, eat too often.[120]

It is this last comment that is particularly interesting given physical culture's impact on ideas of dietary restriction and its advocacy of moderate fasting for women, which, as Ina Zweiniger-Bargielowska has pointed out, was celebrated, 'in contrast with the private suffering of female anorexics'.[121] Physical culture was also largely responsible for the production of striking visual images of young women exercising, wearing loose robes, special gymnastic outfits or swimsuits that played a vital role in shaping ideas of what the modern girl should look like (see Figure 2.3). Under the influences of dress reform and the rise of the New Woman with her uncluttered, streamlined appearance, girls became free from the trimmings and excesses of late Victorian fashion, notably corsets intended to produce an exaggeratedly narrow waist. Historians of fashion, Buckley and Fawcett, have suggested that by the 1890s the ideal female body was larger than that of the 1860s and 1870s and the fashion was for curves.[122] The 'noble contours' of the Venus de Milo with her full waist appeared again and again to illustrate advice literature as a counterpoise to the badly dressed, corseted young woman with a wasp waste and 'artificial lines'.[123] Calorie counting, carefully controlled diets and concern about the boyish figures exemplified by 'flappers', would only start to have a major impact after the First World War, yet in 1895 *The Domestic Magazine* expressed a wider anxiety about the slimming craze, combined with excessive exercising. Quoting an American author who had been alerted to this trend, the article concluded:

Englishwomen are always rather angular and dowdy, but never before have they made such an exhibit in this direction as they have made this year. I have made numerous enquiries about it among my English friends, and it is unquestionably the result of extravagant athletics, coupled with the universal rage now going on in London for excessive thinness. All the society women in the big metropolis are 'banting' or keeping their weight down by one method or another, and a woman whom we could consider unquestionably 'skinny' is looked upon as the perfection of form in Great Britain.[124]

Though advice literature – as well as the growing number of publications on domestic science – was more likely to recommend sensible diets and healthy eating rather than dieting per se, and to provide information on the nutritive value of particular foods and food preparation, it also suggested ways of tackling overweight and underweight. It was preoccupied too with the theme of constipation, a worrying by-product of taking medicines containing iron for anaemia, as well as a trigger of 'auto-intoxication', a state attributed to the bowel becoming full of poisonous waste food matter. Special exercises were devised to stimulate the bowel, and girls were advised to relieve themselves regularly, to drink plenty of water and to eat a diet rich in fruit, vegetables and wholemeal bread.[125] Advice literature, seldom, if ever, referred to anorexia nervosa, which had been labelled as a discrete condition of adolescence in 1873, though Anna Kingsford stressed

> When attenuation is excessive, when the ribs protrude, when the elbows and knees exhibit the shape of the articulation of the joint, and the face wastes, and the shoulder-blades and breastbone show themselves distinctly under the skin, then special medical advice should be sought, for thinness so pronounced as this indicates disease.[126]

While Katharina Vester has suggested that US health guides for women did not include sections on obesity, the term was certainly used in English publications and English girls were castigated for either eating too much or too little, or the wrong kind of food, and both fat and thin girls were ticked off for taking too little exercise.[127] Girls were urged in general to moderate their eating. Over-consumption was a sign of poor character and greed: 'The value of temperance, both with regard to food and drink, is very great. The girl who over-indulges herself with sweets and cakes has to play the penalty of a sluggish complexion and impaired digestion.'[128] Anna Kingsford opened her book with chapters on obesity and leanness. 'Julia' was instructed to abandon her suicidal consumption of bread, potatoes, milk soup and tapioca pudding, and warned not to take any pills for the mitigation of obesity. She was instructed to become an early riser, to take brisk walks and exercise and to consume uncooked fruits, vegetables, white fish, lemon tea, and rusks, and to avoid farinaceous dishes, milk, sweets, pastry, cocoa and alcohol. Kingsford's letter was largely about moderating behaviour and adapting lifestyle, forbidding Julia from lolling in bed and eating 'stray cakes or cups of tea' and substituting this with gymnastic exercise and Turkish baths. Thin 'Psyche' was advised to overcome her nerves and to cultivate 'a fixed habit of placidity, avoiding fret and mental irritability'. She was instructed to eat a rich diet, including porridge, mashed potato, bread and honey, eggs, fish and vegetables, and to take daily exercise, preferably horse riding.[129]

Aside from dietary tips directed towards weight gain, Emma Walker encouraged 'thin girls' to sleep, take vigorous baths, follow special 'development exercises', enjoy fresh air and sunshine, not to fret and to adopt a more philosophical approach to life.[130] Gordon Stables relayed the dangers of overindulging in the story of 'How Fanny Ffisher lost her figure, and found it again (A true story with a bit of a moral in it).' Fanny's increasing portliness led her brother to liken her to a pug dog, while Fanny hankered after corsets to restrain her bulk. Stables reported that he put Fanny on a diet – forbidding bread, pastries and potatoes – and a regime of early to bed and early to rise, with plenty of exercise on the tricycle. After two months Fanny had slimmed down and soon received a proposal of marriage, which was for Stables an excellent result.[131] Many health guides included exercises directed at reducing weight or bulking up skinny girls, or beautifying and streamlining the arms, legs or waist, emphasising the need for persistent application and patience to achieve results.

In the mid-Victorian period a girl's transition to young womanhood was marked by the adoption of adult clothing and hairstyles associated with 'limitations that such clothing place upon vigorous physical activity', especially play.[132] Increasingly, by the 1880s and 1890s clothing was intended to be enabling, whether it was worn in the workplace, school or gymnasium. Specialised clothing was developed for sport and leisure, including Jaeger woollens for outdoor activities such as cycling and a huge variety of gymnastics costumes. Even those not advocating reform clothing or the divided skirt, still recommended practical and comfortable clothing above style and constraint.[133] Emma Walker put the question to her girl readers: 'Have you ever thought how much more sensible your brothers are in respect to their clothing than you are?'[134] Walker advocated 'nature's corset' and in doing so joined a chorus of writers who supported either less constraining corsets or their complete abandonment. The Olsens declared the wasp-waist 'a Deformity', 'a monstrosity' and published an extraordinary image of

Fig. 1.—The Venus de Medici, showing   Fig. 2.—The outline of the body produced by tight lacing.
the natural outline of the body.   *(From a photograph.)*

Figure 2.4   Venus de Medici compared with a body produced by tight-lacing.
Source: From Malcolm Morris (ed.), *The Book of Health* (London, Paris and New York: Cassell, 1883), p. 498 (author's collection).

'a fashion-deformed woman' to emphasise their point.[135] Similarly, a lengthy and comprehensive chapter on clothing, authored by Frederick Treves and published in Morris' *Book of Health*, employed several images of the internal damage resulting from tight-lacing to demonstrate how this could lead to permanent deformity, damage to the internal organs, poor blood circulation, muscular damage and restricted breathing, as part of a meticulous analysis of the failings of a variety of female attire. The chapter also included an image that compared the Venus de Medici with a body produced by tight-lacing[136] (see Figure 2.4).

Annie Burns Smith's *Talk with Girls* also pointed to a series of health risks associated with the crushing of the organs resulting from tight-lacing, causing shortness of breath, heart palpitations, attacks of faintness and constipation. 'Besides a tight-laced woman cannot eat sufficient food to satisfy the needs of her body...she soon becomes pale and anaemic. The muscles of the trunk, being compressed, become weak and flabby, causing spinal curvature and round shoulders.'[137] She also referred to the dirty habit of wearing long, trailing skirts which picked up sputum, containing the bacilli of consumption.[138] The magazine editor and writer on health, Ada Ballin, observed in her introduction to Howard Spicer's 1900 manual *Sports for Girls* that, whereas a decade ago she was compelled to protest against tight-lacing in her public lectures, now it had largely disappeared, 'a fortunate thing that those who go in for sports cannot possibly make it consistent with their enjoyment of these that they should lace tightly'.[139] Elizabeth Sloan Chesser confirmed that 'the spread of knowledge regarding the need of physical exercise and freedom of movement has practically abolished the rigid "stay" corset of the past, and modern corsets are lighter and more flexible, and are constructed on hygienic lines', even if at times they produced 'injurious compression, to interfere with respiration, circulation and digestion'.[140] A section of Mary Humphreys' *Personal Hygiene for Girls* entitled 'Dress and Character' warned girls of 'that love of dress and outward show which becomes a vice in many women...**Your outward expression shows your character.**' She recommended a yoked box-pleated tunic for young girls and for older girls a neatly made shirt blouse, 'perfectly fresh and clean', in combination with a plain skirt; a costume 'suitable for girls of any station'.[141] In 1918 the Stenhouses were still castigating girls who wore corsets, as well as all other tight clothing, shoes and boots, hats, gloves and stockings. They also advised girls to invest in quality garments rather than cheap materials, and criticised the 'slackness' associated with failing to keep clothes in good repair.[142] Preferred fabrics were wool in winter, cotton and linen in the summer. Loose-fitting clothes were recommended and girls were cautioned to cover up to avoid chills: 'To leave the chest bare, or cover it with a thin, transparent blouse, whilst the lower part of the body is heavily clothed, and the feet and legs again are covered in thin stockings and thin-soled shoes, is unhygienic, and, indeed, dangerous to health'[143] (see Figure 2.5).

Health in the *Girl's Own Paper*

The *GOP* first appeared in January 1880, a year after the successful launch of the *Boy's Own Paper*.[144] Given its huge circulation and the wealth of material on health included in its pages, the *GOP* provides

Figure 2.5   'The result of too hasty toilet. Notice the safety pin'.
Source: From Mary Humphreys, *Personal Hygiene for Girls* (London, New York, Toronto and Melbourne: Cassell, 1913), p. 95 (author's collection).

an interesting case study of the form and range of health advice offered to young women, particularly as the paper straddled an uncomfortable position between the idealisation of Victorian girlhood and femininity and promotion of the modern girl. Produced by the Religious Tract Society, the *GOP* was steered by moral and religious imperatives and catered for 'readers standing midway between girlhood and adulthood', typifying the 'New Journalism' separating girls from women readers, although much of its content was also directed at younger girls and older women.[145] During its first 28 years under the editorship of Charles Peters, the *GOP* was confronted with new definitions of young womanhood and femininity, and its writers presented diverse views on the modern girl, while the magazine also boldly announced its appeal 'to girls of all classes'. Peters explained 'Girls of a superior position ... should read everything, and be well up in *every* matter, upon which we give instruction. Their money, time and superior intelligence admit of this.' For girls of less exalted status, 'there are papers on economical cookery, plain needlework, home education, and health', and 'servant girls had written describing the papers as a "companion" in their isolation and a counsellor in times of sore temptation'.[146] Much of the *GOP*'s content was improving and instructional, covering topics such as careers for girls, languages, crafts, nature, sporting opportunities and recreations, dressmaking, cookery, and health and beauty, but also included articles on fashion and etiquette, and fiction and poetry. The magazine was a huge commercial success and rapidly achieved weekly circulation figures of 250,000, outstripping sales of the *Boy's Own Paper*.[147] Given the size of the readership, its potential impact on the topic of health was substantial.

Under what appears to have been a loose editorial regime, the *GOP* was ambiguous in its stance towards young women's changing status and new opportunities, with individual articles reflecting varied viewpoints but much of the fiction 'was reactionary, working to absorb and tame rebellion' and Kimberley Reynolds has described it as 'ultimately conservative'.[148] Penny Tinkler, by contrast, has suggested that the *GOP* was very much in tune with the views of secondary school headmistresses and the ethos of the girls' club movement,[149] and indeed some copy advocated the benefits of girls' clubs and promoted the extension of secondary schooling for girls and women's careers and opportunities, urging a change of focus from domestic concerns to public life. Information was provided on a huge range of careers, including medicine, dispensing, public health, the civil service, teaching, art and architecture, archaeology, farming, gardening and journalism, and many of the articles provided detailed information on courses, the cost of training and admission procedures, as well as lauding the successes of pioneers in these fields.[150] Yet other contributions declared women unsuited to mental exertion and advised against taking 'masculine subjects', and the magazine strongly promoted traditional domestic skills, particularly child care, sewing, cooking and home nursing.[151] Visually, the magazine maintained a staid image, with illustrations depicting young women as ladylike and feminine, lacking in dynamism, decorous and overdressed, even while going about stirring activities such as tennis, cycling and hockey.

In terms of its commentary on health and physique – as well as many other subjects – the magazine 'reflected a new female self-consciousness and revealed the contradictions it involved'.[152] An article published in 1894 and cited at the start of this chapter, idealised the 'working woman' and praised modern girls for their 'strength and purpose'.[153] Less than a year later, however, another feature suggested that the 'modern girl' had improved physically and mentally, had become stronger, more poised and straighter: 'Is she not more active? Can she not run faster, jump higher, endure longer? Is not her skin clearer, her eye brighter, her hue healthier?' Yet morally and spiritually, improvement was dubious, the article went on, and physical and mental betterment had its downside, with the modern girl in danger of becoming physically 'overpowering and wanting in repose', 'priggish' and 'morally loosing her sweeter traits and sometimes hard and selfish'.[154] The mocking of 'Some Types of Girl-hood or Our Juvenile Spinsters' took as cruel a turn as that of the feeble, sofa-bound girls of the mid-Victorian period, with the defining of liter-ary, artistic, common-place and muscular types; dangerous manly sports were 'annexed from the exclusive proprietorship of the stronger sex', resulting in masculine postures and 'the phraseology of the race-course and stable, the barrack or the cockpit'.[155] In the same year, another writer described how 'the girls of today' were liable to be worn down by hurry and anxiety. 'Perfect physically, full of the abounding joy in life which health and true carelessness brings, . . . they are apt, in their very exuberance of life and spirit, to ride over and push to one side the feelings of those who humbly wait upon them.'[156]

The *GOP* tackled issues dealing with the practical and personal prob-lems confronting its readers in their daily lives, exemplified in its 'Answers to Correspondents' sections (questions were not printed). These demonstrated the extent of interest in health matters and the pragmatic approach of respondents to anxieties about wellness and fitness.[157] Health came up frequently too in tips to readers and remarks appearing under the heading 'Varieties'. On 'Women's Health', for example, 'Women have themselves to blame for the greater part of their weakness and diseases; it is caused by their foolish and unhealthy habits than anything else [*sic*].'[158] 'A Good Word for Tight-Lacing' praised it as a public benefit, 'because it kills off the foolish girls and leaves the wise ones to grow into women'.[159] Corsets and their evils proved to be an endless source of interest in correspondence pages. In a reply to 'Two Anxious Readers' in 1888 the inquirers were warned of the multiple dangers of tight-lacing, which 'produces redness of the nose and disease of the heart. Besides this, it may cut through the liver, or lungs, and destroy the digestive powers. It impedes the circulation of the blood, and tends to produce chilblains. Besides this, it ruins the personal appearance.'[160] Fair Japan was advised that it would do her no harm to ride a bicycle in 1898, provided she sat well on the machine and did not ride too fast. The same issue offered tips on bronchitis, catarrh of the throat and indigestion, and an inquiry from 'Arthur' advised an ointment for acne and recommended that tea be given up entirely.[161] A reply to the 'Home Girl' in 1899 took a grounded approach to hysteria:

> The 'fits' you describe are typically hysterical…Hysteria is a serious malady, but it is curable, not by the physician but by the patient. You must overcome the feelings which trouble you. We know it is hard to do so, but mental restraint is the great curative agent for hysteria. If you can live an outdoor life, you should do so; healthy open air employment is far less likely to produce hysterical fits than working in a cramped position in a badly ventilated room. Eat well, but take no stimulants and no medicines. A morning bath and very mild gymnastics may help to strengthen your nervous system. And above all things beware of sympathising friends. Sympathy is all very well in its way, but it can do more harm than anything else. Sympathy may be the cause and the object of hysteria.[162]

Anonymous though these correspondents were, they testify to enduring interest in health matters. Meanwhile, the replies, firm and not open to contradiction or negotiation, offered straightforward advice with a problem-solving emphasis which apparently drew little on what were very influential medical theories emphasising female weakness and liability to ill health, particularly around puberty.

For almost 30 years, Gordon Stables, alias 'Medicus', hosted the *GOP*'s health column, producing regular features on health maintenance, exercise, sport and fitness, diet, bathing and beauty regimes, the dangers of tight-lacing, the treatment of minor ailments and home nursing, and cornering something of a niche market in writing 'popular medical literature for magazines and journals patronised by the fair sex'.[163] His books and articles in the *GOP* were quirky, chatty and packed with homilies, and emphasised a straightforward approach to health, based on exercise, healthy food, avoidance of alcohol, cleanliness, sufficient sleep and mental stimulus. Medicines were of dubious value, 'folly' even. He recommended athletics and cycling to his readers, claiming to be one of the pioneers of cycling, and advocating it as a cure for various chronic ailments, as well as for digestive disorders, sleeplessness, nervousness, depression and overweight.

Stables' advice was predominantly of the straightforward, bracing no-nonsense variety, drawing in all likelihood on his background as a naval surgeon, cold baths, hot porridge, brisk exercise and early to bed. He appears to have engaged little with medical theories on the risks to body and mind faced by adolescent girls, and his straightforward approach urged strengthening regimes on his readers to boost their energy and application to a whole range of tasks. However, while he encouraged a great deal of healthful activity, it was less clear what Stables actually expected girls to *do*, aside from bustle around, take on charity work and sick nursing, enjoy games and fun, although he acknowledged that some girls would be obliged to work for a living. Though those following his advice would benefit in their good health from enhanced enjoyment of life and improved spirits, 'too much work and overstudy are both sure to weaken the body and prepare it for the reception of any imperfection or passing ailment'.[164] Certain topics were not discussed by Stables or other *GOP* authors – menstruation, sexuality and childbirth – and neither was the question of girl's future potential as mothers, though Stables' ultimate goal for girls was marriage to equally healthy young men and motherhood. He concluded that the majority of the readership of *GOP* were 'so healthy in lungs and nerves, and so stout-hearted and strong-limbed, that it is, as a rule, a matter of entire indifference to them how the wind blows or how the weather is'.[165] Writing on how girls could increase their strength, he provided an A to Z of dietary hints, and advised cold morning baths and walking: 'Do not stop to stare in shop windows, but walk as if you meant doing something [*sic*].' After building up to several miles a day, girls were ready to launch themselves into a gymnastic programme.[166] Advocating cold bathing in 'A New Year's Health Sermon':

> Now prey don't misunderstand me; there are some girls whose hearts are too feeble to admit of their plunging into ice-cold water, or using the sponge bath. They may have a dash of hot water in it. Happy are those, however, who can take it cold. I believe they are sure to get married sooner than the others. They will, if they are careful in diet, exercise (recreative), etc., soon become hale, healthy and happy.[167]

Stables' girls were encouraged to be fit and strong; 'the muscles should be hard and firm', maintained by gymnasium work and callisthenics.[168]

He likened views that girls need not exercise and should stand by and watch 'their brothers or friends show off their prowess on the field, or in the gymnasium' to fairy stories: 'But those days have gone by, never to return, because we know now that there can be neither health nor perfect happiness in a feeble or badly developed body or frame.'[169] Good character went hand in hand with good health, according to Stables, who demanded a certain amount of pluckiness but also sweet-temper from his readers as well as a regard for correct principles as these applied to behaviour and rules of health. 'Keep a smiling face...fretting weakens the body.'[170] 'The mind has much to do with the health of the body. Try to control your temper, never get angry,...Read good books, especially religious books'.[171]

Gordon Stables subscribed to a straightforward theory of energy supply, at odds with the 'fixed fund of energy' thesis and the pronouncements of many of his professional colleagues described in the previous chapter. He advised that some girls had the potential for more improvement than others, with 'depressive and nervous girls', that is, middle-class girls with no employment outside the home, being cautioned to be careful in their approach to exercise.[172] Working girls were depicted as more robust, yet they were also warned – despite the economic imperative to work – not to overexert themselves and to take proper nourishment, rest and outdoor exercise.[173] Stables recognised the scope for incremental improvement for all young women: 'All kinds of exercise do good; walking for the weakly, cycling and rowing for the stronger, the dumb-bells and Indian clubs before breakfast or in the afternoon for all.'[174] The *GOP* also devoted considerable attention to 'invalid' girls and chronic cases, those who were 'never overwell', tired, subject to low spirits, aches and pains, and who had little inclination for exercise. Stables suggested that many of these girls relied too heavily on medicines, lacked occupation and that their 'minds might be said to be preying on their constitutions'. The solution for this was to replace medicines with fruit and vegetables, and to grow them themselves, rising early in the morning and working an hour before breakfast, and doing a little more work in the evening. This would result in purer blood, stronger nerves and a happier state of body and mind.[175]

While Stables was the most prolific and enduring of the *GOP*'s health writers, the magazine regularly published material by other contributors on health, exercise, deportment and beauty. Mrs Wallace Arnold's 1884 article meticulously outlined an exercise regime for girls, and urged careful balance between higher education and physical development, recommending supervised callisthenic exercises.[176] In the 1890s many articles reported on and advocated sports for girls, especially cycling, but also hockey, rowing, golf, croquet, archery, tennis, cricket and a variety of outdoor games. The *GOP* published a series of articles on the 'Physical Training of Girls' in 1900 and Stables himself adopted this term in 1899, though he perceived physical culture as beneficial in terms of its emphasis on regularity but not rounded enough to result in true health.[177] In 1897 the 'New Doctor', Dr Lawrence Liston, joined the *GOP* team of contributors, indicating that the *GOP* perceived that there was an enduring and far from satiated interest in health matters. Forging his own position in the magazine, writing on topical medical issues, including cosmetic medicine, he also overlapped with many of Gordon Stable's pet topics, such as nervousness and how to banish it, holidays, cycling for health and sea bathing. However, Liston urged, to an apparently greater extent than Stables, moderation and care in taking up exercise and new regimes of health.[178]

Conclusion

There were clearly limits to the expectations of Stables and other *GOP* writers on what enhanced health would enable girls to achieve and indeed how far they could actually advance their health. More generally, the magazine was ambivalent about what its girl readers would and should be able to do in the worlds of work, education or recreation. The *GOP* published on, and generally lauded, pioneering girls' schools and colleges and declared exercise an excellent counter to the demands of study; 'one of the most beneficial results of a really good education is undoubtedly the equilibrium established between the respective powers, mental and physical', Mrs Wallace Arnold opined in 1884.[179] Yet she then went on to argue that girls had an additional imperative in training the body: 'Surely the future wives and mothers of England – for such is our girls' destiny – may lay claim to no less share of attention.'[180] By the early twentieth century, the magazine celebrated – in a similar way to the authors cited at the start of this chapter – doing away with the negative image of Victorian girls and even earlier stereotypes. 'Do girls no longer faint? ... Look at a bunch of modern girls in the open, at their sports or fun, and the truth is driven home to one that we have travelled a long way from the swooning young lady of Jane Austen!' Such a girl 'seems to be disappearing, and she seems to have disappeared through the open atmosphere of the girls' sports, and better self-control all down the line'.[181]

Writers on girls' health more generally urged a dynamic approach to health which had a purpose beyond producing well-honed bodies and

physical and mental wellbeing. A strong moral imperative also permeated the information presented to girls – it was largely though their
behaviour and self-control in adhering to the rules of health that they
would achieve their goals of becoming fit but not manly, shapely but
not plump, strong-minded but not selfish, cheerful, lively and energetic
but not boisterous or cocksure. The change in girls' visible, outward
appearance between the late Victorian period and the early twentieth
century was little short of stunning, notably in their dress but also, many
contemporaries noted, in their physical stature, strength and energy levels. The information contained in advice literature liberated girls from
the idea that they were victims of their biological bodies, and assured
them that they had the ability and wherewithal to achieve robust health
and, indeed, that this was their responsibility. Energy was presented
not as a limited resource, but as something to be managed and even
enhanced. This implied a reframing – or rather a sidelining – of the
medical literature examined in Chapter 1, which stressed the dangers
of puberty and the manifold risks to health for adolescent girls, and
implied that girls would have few resources to tackle their poor health
themselves.

Yet girls were not accorded authority in managing menstruation, an
issue vital to them in understanding their own bodies, and Kirk's more
detailed – even if inaccurate – advice on the facts of life remained a rar-
ity. Moreover, it was only during the First World War that the subject of
sexual hygiene was broached more regularly, as the apparent slackening
of public and private morality and the release of large numbers of young
women into the workforce heightened anxieties about the dangers of
venereal diseases and the abandonment of young fallen women, themes
taken up particularly by eugenicists such as Mary Scharlieb and organisations such as the Young Women's Christian Association. Coverage of
sexual hygiene was, however, boosted by the physical culture movement
and the importing of a literature, notably from America, which tended
to be far franker on such issues. In 1918, the Australian physical cultur-
alist Annette Kellerman linked bodily beauty with sexual attraction, a
right and good association in her view, and emphasised the role of the
wife as a sexual partner, rather than mother, placing value on youth and
women's responsibility for their own lives and winning a husband.[182]
This was a startling turnaround to viewpoints expressed by authorities as
diverse as Victorian physician Lionel Weatherly and eugenicist authors
who cautioned mothers and girls about inappropriate relationships as
girls evolved into young women. It was also in stark contrast to the
literature produced on falling moral standards and the 'selfishness' and
'love of ease' associated with the declining birth rate. In an environment

of apparently loosening moral standards, Mary Scharlieb warned too of the effects of venereal diseases on adolescent girls of all social classes:

> The mother consults a doctor, and asks: 'What can have happened to this girl? She was such a clever, such a fine girl...she could play hockey and tennis, she was good at the piano, she was good at languages.' And another 'I had great hopes of my girl; I thought that she would go to Girton or some college...Now her memory has failed, and she has all sorts of queer fads and fancies...she cannot eat'.[183]

Girls would be continually warned, too, about the dangers of alcohol, narcotics and tobacco. Those who succumbed 'were on a road, a steep road, and behind a runaway horse'.[184] 'Apart from the direct effect upon the organs of digestion, the liver, the brain, and the nervous system generally', Chesser described intemperance as 'a habit which causes serious deterioration in character'.[185]

The impact of advice literature, in terms of who read it and who acted upon it, is difficult to assess, though the apparent marketability of girls' magazines and other prescriptive material 'surely spoke to a felt need of the readership'.[186] Publishers continued to bring out new and updated issues of advice books, indicating their enduring popularity, and the subject was accorded a vast amount of coverage in magazines and periodicals. Girls were offered instructions, information and admonitions on health from many directions – in the magazines they purchased, in domestic literature to be found at home, on the bookshelves at school or club libraries, in lectures at the workplace. We do not know, however, how many girls read the literature or took the advice on board – we can only suggest that they were exposed extensively to it, though they certainly corresponded vigorously with magazines like the *GOP* on health matters. Certainly, many more individuals and organisations had a stake in producing information on girls' health, which was viewed and commented on in diverse public arenas and which, as the following three chapters will illuminate, became more visible in public and semi-public spaces.

The impact of much of this advice literature during the closing decades of the nineteenth century and early years of the twentieth was to demystify health. At the same time, the need to produce good health was regarded as pressing. Health became inexorably linked with progression and with self-management, it was aspirational and fashionable, as well as necessary for the broader public and national good. Recommended regimens were time consuming and put considerable pressure on girls who were urged to take control of their bodies, which in many ways they still had only a partial understanding of. Amy Barnard warned in 1912 of the dangers of over-doing physical activity, pointing out girls' susceptibility to a range of ailments, and urging them to look out for warning signs of sickness and disease – over-heating during a game of lawn tennis or hockey could prove lethal; so too could remaining too long in the sea, or wearing a 'pneumonia blouse'.

> And, yet again, here is a girl of seventeen who overstrained the focus of her eyes by excessive study, and in consequence has to wear spectacles. At the school I used to attend was a young boarder who went skating after discarding an undergarment, and paid for the act with her life... During the summer of 1911, two girls of whom I know paid heavily for disregarding their bodies. One went gardening with the full blaze of a hot summer sun upon her head, and in a few days died in convulsions; the other remained in the sea with the sun's rays beating on her head till she had a fit, and nearly lost her life. The mention of these dreadful incidents will serve its purpose if, on reading them, you realise the duty of caring for the body, and of avoiding crippling it or destroying its usefulness.[187]

Diet, sleep, exercise, washing and dress were to be closely monitored, timetabled and controlled: 'Besides attention to regular habits, sleep, ablutions, fresh air, physical culture, and rest, there are food and clothing to be considered.'[188] Regimes such as Barnard's imposed – along with constant watching for signs of illness – great stress on readers if they attempted to follow this advice to a letter, particularly when their behaviour and emotions potentially had such a strong impact on their wellbeing. Barnard certainly represented an extreme view but, as the examples in this chapter have illuminated, girls were urged to take on great responsibility in terms of health, what they ate, how they kept themselves clean, what they wore, and how they regulated their thoughts and activities. In this sense, girls' behaviour was presented as liable to expose them to as much risk of ill health as their biological vulnerability.

3
Health, Exercise and the Emergence of the Modern Girl

In 1899 Dr Arabella Kenealy, physician, author and 'eugenic-feminist', regaled the readership of the literary magazine, *The Nineteenth Century*, with the story of Clara: 'A year ago', Kenealy wrote, 'Clara could not walk more than two miles without tiring; now she can play tennis or hockey, or can bicycle all day without feeling it'.[1] Through exercise and a vigorous lifestyle, Clara had toned and honed muscles, was slimmer, stronger and more agile. However, she had bartered many qualities for a 'mess of muscle' and lost her subtle charms – sympathy, patience and an elusive beauty – in the process of developing an athletic body.

> Clara the athlete was no longer the Clara I remembered two years earlier. She was almost as dissimilar as though she had been another personality... She was then – she is now – something more than comely, but her comeliness has altogether changed in character. Where before her beauty was suggestive and elusive, now it is defined... the haze, the elusiveness, the subtle suggestion of the face are gone; it is the landscape without atmosphere.[2]

In place of this Clara had produced a highly toned body, briskness, mere muscular achievement and a 'bicycle face (the face of muscular tension)'. She had a loud strident voice, neglected her domestic tasks and her siblings, and had abandoned gliding in favour of manly striding.[3] Worst of all, Clara's pursuits were endangering her potential as a future mother; 'it is the birthright of the babies Clara and her sister athletes are squandering'.[4]

In contrast to 'Lawn Tennis Girl', lauded by gynaecologist and nerve doctor William Smout Playfair, 'bicycle face girl' embodied an exceedingly negative vision of the dangers of overdoing sporting activity and fixating on physical development.[5] As cycling took off as a craze for women and girls, it also became the focus of a medical and moral panic, with one contributor to the *Practitioner* exclaiming in 1895 that 'It would be deplorable if girls whom lawn-tennis has made, like daughters of the gods, divinely tall should be made humpbacked by the bicycle!'[6] Clara was a much touted stereotype in periodical literature, emblematic of the risks to femininity and female body norms which accompanied vigorous exercise.[7] Kenealy was not, however, without her critics. Her article drew an immediate response from the writer, social reformer and feminist Laura Ormiston Chant, who argued in a response in *The Nineteenth Century* that female athleticism embellished rather than threatened the potential of the race, and that muscular vigour and moral qualities (including overcoming criticism such as Kenealy's) would create a nobler and happier home life for athletic girls. In the meantime, Clara, Chant concluded, was likely to still be attentive to her family duties, having become more energetic, more efficient and more fulfilled through her cycling endeavours.[8]

Having the final say, Kenealy came back on Chant's reply, claiming that the stress of 'over-athletics' was likely to lead, referring to one of her preoccupations, to de-sexing; 'masculine women and effeminate men – neuters – spoiled copies of the human edition'. Clara, believing that she was becoming stronger and stronger, was in fact 'converting womanhood into mannishness by the artificial stimulation of the masculine strain in her'.[9] The bicycle, meanwhile, 'by reason of the exhilaration and excitement attending its use – [was] most dangerously prone to convert itself into a hobby-horse which rides its master (more still its mistress) to destruction (physical). And in the hands of growing and misguided persons it assumes the quality of a menace.' Physical exercise was liable to produce, Kenealy continued, alongside general degeneration of the organs, cancer in women and gout in young girls. Both these conditions, along with tuberculosis, lunacy and epilepsy, were on the increase, and the only remedy was 'wholly and absolutely the conservation of womanly forces. The woman whose physical completeness precludes her from spending all her energies in muscular or mental effort stores these for her children.'[10]

It is, in some ways, disingenuous to employ this striking example. Kenealy was amongst the most strident and vocal of eugenicist commentators, whose interests in sexuality and marriage were dominated by anxiety about the future of the British race rather than feminist concerns. She likened the passage of girls through adolescence to a phase of invalidity, an arrested state en route to womanhood, during which time, 'the curving bones, the expanding pelvis, the rounded contours, the inhibited muscles, the languors and recurring disabilities, are designed to restrict activity, physical and mental'.[11] Her views were not widely shared by medically qualified eugenicists and certainly not by the medical profession more generally, many of whom argued that moderate exercise would destroy neither girls' feminine physique and appearance nor their characters. However, Clara's story connected to a set of much broader concerns – albeit usually expressed less fervently – about the dangers young women faced in pushing sporting endeavour too far, transforming themselves into masculine types. As James Whorton has suggested, 'the distaste for women following mannish pursuits was publicly shared by more than one physician. Shadwell, discoverer of bicycle face, deplored the fair sex's fondness for "unsexing itself"; another observed that not only had the wheel "displaced the horse, [but] in women has, in a measure, replaced the uterus." '[12]

These kinds of assertion – of the potential for exercise to go wrong or too far, resulting in the development of a hard masculinity, an overly athletic body, and a potential deterioration in character and maternal function – peppered debates about exercise and athletic performance for girls more generally during the late nineteenth century. This coincided with the increased access of young women to a wide range of modern sports following on from – and no doubt to a certain extent emulating – the obsession with health, sport and physical prowess and, as Haley has argued, spiritual wellbeing, that transformed Englishmen and boys into athletes and team players between 1850 and 1880.[13] These debates featured extensively in the medical press, notably the pages of the *British Medical Journal* (*BMJ*), as well as a wide range of popular health, literary and general interest periodicals, girls' magazines and advice books.[14] *The Nineteenth Century*, for example, published numerous articles on exercise and athletics in general and cycling in particular around the time that Kenealy published her account of Clara and her bicycle face, authored by both doctors and lay commentators. Though many periodicals presented a range of opinions on the pluses and minuses of girls' sporting endeavours, as interest in the New Woman peaked in the 1890s, they became engrossed in the subject of 'Manly Maidens': girls seeking to earn the 'proud title of "a good fellow" by emulating the fashions and the habits of the robuster sex' and indulging in 'hygienic excess' and inappropriate sports.[15] Those promoting exercise and gymnastics activity, meanwhile, enjoyed considerable access to print media. The pioneer of physical education for girls, Madame Bergman Österberg had tributes to her work appear in the *Lancet, Women's Herald, Ladies' Pictorial,*

Pall Mall Gazette and numerous other magazines and journals, while early advocates of physical culture, notably Eugen Sandow, enjoyed enormous publicity and press exposure.[16]

This chapter surveys turn-of-the-century debates on the relationship between health, exercise and sport for young women, focusing specifically on the example of cycling. As medical debates on the pros and cons of exercise and sport for girls spilled over into popular media, including magazines and books catering specifically for girls, this became crucial in providing information and advice on how to take up such activities, as well as actually promoting or critiquing them. And while many sporting endeavours were confined to the well-to-do, by the close of the nineteenth century working-class girls also participated in exercise regimes and a range of sports in significant numbers, even if their choice of activities was much more constrained. Some sports, meanwhile, including cycling, became less costly and thus more widely available. Cycling – though a site of medical contestation – was promoted by many physicians, not least because they often cycled themselves. It was depicted as an ideal form of exercise, a potential pathway to health, and it was even lauded as a therapeutic agency for specific disorders and conditions. Physical activity and outdoor sports also became embedded in broader regimens of health, with many doctors spelling out a course of action for girls, intended to improve their wellbeing, involving exercise, hygienic practices, nutritious diet and appropriate dress, and their uptake was associated more broadly with interest in the production of healthy bodies and minds. As with other healthful activities, exercise was linked inexorably to girls' behaviour and habits; if practised in moderation, it could counter the limits imposed by their biological make-up and work to produce more vitality and energy; failure to moderate could, by contrast, have disastrous consequences.

The moderation of exercise for health

From the mid-nineteenth century onwards, feminist support of physical activity was important in driving a change in attitude towards the participation of girls in sport and in overcoming moral and medical objections to this. Sport was also placed firmly on the agenda of girls' schools, as we will see in the next chapter. Feminists and, later, New Woman enthusiasts supported recreation that linked girls to the outdoor environment, encouraged mobility and built strength, as well as, through the gradual abandonment of restraining dress and of chaperonage, bodily and spatial freedom. As early as 1850, Harriet Martineau had recommended activities such as swimming and rowing to girls, while the founder of the *English Woman's Journal*, feminist Bessie Rayner Parkes, campaigned for better provision of physical training in girls' schools, regretting that 'most people endeavoured to check the physical power of their daughters as much as that of their minds'.[17] A number of periodicals charted the progress of female swimming clubs and campaigns for women's hours at public pools, likening women's exclusion from such activities to the restrictions of jealously guarded men's clubs.[18]

Catherine Horwood has suggested that 'the study of women's leisure patterns is a vital barometer of their ability to achieve equality both within and outside the constraints of family life', while Sheila Fletcher and Kathleen McCrone have described the New Woman's indulgence in physical activities and the take up by women and girls of competitive sport as challenging male exclusivity and representing feminist hopes and ambitions.[19] The archetypal image of the New Woman is indeed an energetic, neat and confident young woman astride a bicycle. For most women and girls, however, it is likely that their exercise and sporting ambitions were achieved as part of more modest agendas, the desire to work in the gymnasium or to play sport at school, to take on similar sports to their brothers, or to enjoy the daily practice of exercises. It is also likely that their ambitions were stirred and spurred on by the material they read, including novels of the period, with their sporting heroines, and periodical and magazine articles advocating healthful exercise.[20] As one contributor to the periodical *Health* suggested in 1895, the fact that girls had successfully gained access to a wide range of sports – golf, tennis, riding, boating and outdoor sports and pastimes – was symptomatic of the changing landscape of opportunities for young women.[21] Exercise was promoted in a multiplicity of arenas; aside from girls' schools, in gymnasia, sporting clubs and girls' clubs, in swimming pools and at the seaside, in a small number of factories and other workplaces, and, in the case of bicycles, public highways and parks. In 1899 the *Girl's Realm* magazine boldly claimed that not only was the gym essential at girls' schools; in private households 'modern girls' expected to pay weekly visits to the gym, suggesting that the take up of such activity was becoming fashionable and desirable for those with the wherewithal to afford such pursuits.[22] Additionally, much advice literature advocated daily regimes of gymnastics and breathing exercises designed to improve tone and physique, which were designed to be carried out at home.

Numerous role models promoted the outdoors and physical activity to girls, notably with regard to cycling, which became popular with

headmistresses and 'cycling heroines', including members of the aristoc-
racy and well-known actresses. Miss Hope Temple, author of 'pop songs'
and admirer of the New Woman, reported to *Good Health* in 1895 that
she expected to compose new songs on her bicycle. She was 'in love'
with her machine, despite having come to grief riding downhill on her
second outing. She wore a short skirt and sailor's hat to cycle, 'just the
same costume as one has for glacier climbing', she casually remarked.
She declared herself fond of many other sports and did not see why
women should be debarred from any if they felt 'physically fitted' for
them. For her, the New Woman was typified by one of her friends –
highly educated, 'but at the same time she is strong and healthy; and
adept in all kinds of athletics'. She was also 'womanly to the core, and
has all the charms of a pretty English girl. Such will be, in my opinion,
the woman of the future.'[23] The *Lady Cyclist* ran a series entitled 'Lady
Cyclists at Home' and the *Hub* one on 'Women of Note in the Cycling
World', based on interviews with semi-professional cyclists, actresses
and society ladies. These accounts espoused physical strength and ability, stamina, proficiency, agility, love of the outdoors and daring to their
readers.

The range of sports taken up by girls expanded rapidly in the last quarter of the century, even though many of them were medically disputed
or opposed on moral grounds, or both. In 1889 Dr Pye Henry Chavasse's
durable and popular health manual promoted a number of sporting
activities for girls who were too often, in his view, stuck indoors. Many
built character as well as fitness and form:

> Archery expands the chest, throws back the shoulders, thus improv-
> ing the figure, and develops the muscles... Horse exercise is splendid
> for a girl; it improves the figure amazingly – it is most exhilarat-
> ing and amusing; moreover, it gives her courage and makes her self
> reliant. Croquet develops and improves the muscles of the arms,
> beautifies the complexion, strengthens the back, and throws out the
> chest... Croquet has improved both the health and the happiness of
> womankind more than any game before invented... Croquet brings
> the intellect as well as the muscles into play.[24]

A decade later, Mary Whitley's *Every Girl's Book of Sport, Occupation and
Pastime* offered instruction on horse riding, cycling, rowing, swimming,
skating, golf, lawn tennis, croquet, cricket, badminton, ball games and
archery.[25] By 1905 Dr Emma Walker's *Beauty Through Hygiene*, alongside a
detailed regime of exercise and breathing techniques to be carried out at

home, approved golf, rowing, canoeing, lawn tennis, swimming, bowling, skating, hockey, fencing, basketball and horse riding (side saddle to avoid the 'considerable injury' of using a man's saddle) and gardening as appropriate for girls. Bikes did not feature, which is not surprising given the fact that Arabella Kenealy edited and introduced Walker's volume.[26]

To varying degrees, a number of individual sports were criticised as being potentially harmful and in 1894 one 'celebrated medical man' argued that walking was the only exercise appropriate for girls. Tennis was too violent, likely to lengthen the arms and make the shoulders uneven, cycling made women awkward and caused weakness of the back, and cricket, riding and golf also made women 'uneven'. Women were also likely – more likely than men apparently – to be hurt by tennis balls, to fall off horses or tricycles, or be hit on the head by golf clubs. While unusually obsessed with women's proneness to accidents, this anonymous doctor represented a much vaunted view of the dangers of physical deformity or over-development of particular muscle groups that could follow on from female sporting activity, although some physicians pointed out that apparently harmless domestic pursuits, such as piano-playing, fancy needlework or sketching, could also result in overexertion and deformity.[27]

Many debates focused more broadly on girls' ability to take on physical activity. These were neither governed by a straightforward medical discourse based on assumptions about the fragility and poor biological make-up of women who 'possessed only a limited amount of energy' nor typified by notions 'that adolescent girls should take plenty of rest, avoid intellectual overstrain and take moderate exercise in the open air', though many physicians applauded these concepts.[28] But while some commentators focused on girls' potentially meagre supplies of energy, others expressed concern at their lack of physical activity, and a great deal of medical and popular literature devoted itself to the task of urging girls to be more active and those in charge of them – parents, schools and other agencies – were encouraged to adopt initiatives to facilitate this. In 1879 J. Hamilton Fletcher cautioned in *Good Words* against neglecting girls' physical wellbeing, suggesting that 'the weakness and positive ill-health which is the lot of so many girls . . . is one of the greatest hindrances to their progress, and it arises almost entirely from the prevailing want of attention to physical training'. Parents, she argued, took 'but little heed of the muscular education of their girls', while the 'creature with limp, powerless arms and a wasp-like waist constitutes the popular idea of female beauty'. To facilitate the active forming of girls' minds and their education 'the physical training of girls must be carefully considered, and their constitutions strengthened and per-fected by proper food, proper clothing, and proper exercise'. Through her observation of girls in the classroom, Fletcher concluded that fit-ness was vital to stimulate intellectual development, just as it was for boys. She advocated gymnastic exercises for every school curriculum, brisk walks, swimming, skating and rowing; the latter was excellent for the development of the arms and chest and would help combat consumption and asthma. Such a programme of activity, she argued, would abolish high shoulders, narrow chests, poor appetites, ill health and mental dormancy amongst young women, as well as nerves and hysterics.[29]

A decade later E.D. Bourne, author of a teacher's manual of girls' games, suggested that, while there had been improvement in girls' edu-cation, particularly for the daughters of the rich and the artisan class, far more attention was still paid to boys' training, notably their out-of-school development. Men's pursuits tended to be healthier and afford more opportunities for preserving health. This was 'unfortunate'

> for women's adult lives are necessarily so much more contracted than men's, whether they become mothers of families, and are absorbed in domestic duties, or whether they prefer or are compelled to be bread-winners for themselves, that there is a special need of secur-ing for them while they are young as much physical exercise and strengthening of their muscles and limbs as can be ... It is extremely important, therefore, that, in the years of training and preparation for their future careers, girls should not only be provided with as large a stock of health as possible, both mental and physical ... Girls' games, no less than girls' school lessons, should be such as are best fitted to qualify them for the strain and struggle of adult life.[30]

In 1891 Dr Alfred Schofield concluded that young women still took far too little exercise and played far too few games, which potentially undermined their physical wellbeing. Most notable was their failure to develop lung power, 'solely from lack of physical exercise'. He sit-uated this for both boys and girls in the context of the increased stress associated with urban life and lack of opportunity for rigorous exercise:

> the tremendous straining and training of the mind that goes on, and the decreasing amount of manual labour, and even walking exercise, from the increase in trams, trains, buses, etc., and we confess that

we consider these outdoor games and exercises necessities of almost priceless value to the population.[31]

Dr Mary Taylor Bissell, who wrote extensively on physical education and household hygiene, explained in a feature that first appeared in the US journal *Popular Science Monthly* and then in *Health* in 1895, how 'the spirit of physical recreation has invaded the atmosphere of the girl's life', considering this in relation to a broadening range of activities, particularly for city girls.[32] Bissell described exercise as vital in building muscle and lung capacity as well as in overcoming physical weaknesses, notably narrow chests and unsymmetrical bodies, disease and functional complaints, such as headache, backache, dyspepsia and neuralgia. She particularly urged that, in addition to gymnastics exercise, girls take advantage of 'every out-of-door sport that she can suitably undertake... both in the sense of opportunity and also in that of consenting public opinion', though in terms of practicalities she considered tennis and cycling most suitable to urban life. She also advised that girls with 'serious curvature' needed to be given specific exercise regimes supervised in the doctor's office.[33] Dr Lawrence Prince noted in the same year how proper exercise could prevent or modify deformity in young women 'by countering the effects of occupation, and may be of heredity. It will thus prevent or modify stoop of the shoulders, spinal curvature, flat chests, etc.' Additionally, 'properly regulated exercise will prevent the excessive accumulation of fat in those predisposed to this condition.'[34]

Sport and exercise was increasingly framed as an investment, a means of banking physical and mental credit, as well as a way of combating stoops, unevenness and other deformities. This would ideally achieve the 'double gain' of more alert, intelligent and engaged young women, who would go on, through building a 'stock' of energy, to become superior wives and mothers, though many would caution, as did Alfred Schofield, that there were risks of injury 'when severe brain-work and very active games are indulged in together; sometimes the double strain is too much'. He suggested, however, that this was the case for both sexes.[35] Girls' physique, he went on, was also of 'national importance to our race'. 'There is no sight that speaks more for the future welfare of England than a group of well-made English girls returning from a tennis lawn; their every movement instinct with healthy life and vigour.'[36]

In 1899, the same year that Kenealy published her denouncement of Clara, educator and advocate of university education for women, Ernest Lowe, criticised the 'againsts', who attacked all exercise for girls, pointing to ' "the bicycle face," the angular figure, the strained, weary expression, the awkward gait'.

> They tell us that athletic exercise is not only physically harmful to a girl, and tends to make her mannish and awkward, and unfitted for the duties of motherhood, but that it has a very deleterious effect upon her mental, moral, and spiritual nature, and if persisted in will inevitably result in coarsening her nature and destroying all the qualities which have ever been woman's chiefest pride and charm![37]

Lowe disputed this, stressing that exercise was a natural process and that girls should start to exercise early just like their brothers:

> The benefits conferred by it upon the modern young woman of the middle and upper classes are almost incalculable... The young women of to-day are finer to look at, straighter, taller, more wholesome-looking, than were those of thirty years ago... The girl who formerly was lackadaisical and languid – never absolutely ill, per-haps, but at the same time never entirely well, always suffering from some trifling ailment, which made her and every one with whom she came into contact miserable – becomes literally a 'new woman'.[38]

Lowe associated physical improvement very much with well-to-do young women, and, while much advice and commentary was intended to bridge the social classes, it was clear that not all would share in the benefits of new sporting endeavours. Martina Bergman (later Madame Bergman Österberg, one of the founders of physical training in Britain), was appointed in 1881 to the London School Board to train women teachers in Swedish gymnastics intended for poor elementary school pupils.[39] Reflecting back on this period in an interview in the *Woman's Herald* in 1891, she commented how she had been 'forcibly struck' by the contrast between 'strong, healthy, well-developed and happy girls, in full enjoyment of all that makes life beautiful... when compared with the ordinary stunted woman forced by unfortunate circumstances to struggle for existence'.[40] By 1891 Bergman Österberg had resigned from the London School Board, disillusioned by the poor physiques of working-class schoolchildren impaired by neglect, and poor diet and living conditions, and with little potential to benefit from physical activ-ities. Undoubtedly, while a wide range of sports opened up to young women during the late Victorian period, many of these were restricted to well-to-do girls who could afford horse riding or access to tennis courts or croquet lawns. For many poor girls, the environment of late Victorian towns and cities offered few possibilities to indulge in sport and exercise. Bim Andrews, an illegitimate girl, born in 1909 and raised in poverty in Cambridge, noted how, as a scholarship girl at the Higher Grade School, she was looking in on the very different lifestyles and leisure activities of fee paying pupils, mostly tradesmen's daughters: 'Long fine stockings, dainty lunches, bicycles, tennis racquets, parties, summer holidays by the sea. I wanted to play tennis, but I had no access to the gear.'[41]

Much advice literature and features in girls' magazines and annuals, meanwhile, circumnavigated the predicament of working-class girls, though health columnist to the *Girl's Own Paper* (*GOP*), Dr Gordon Stables, referred to working girls' need for healthful exercise, pointing out that health was a vital resource for girls earning their own living. He asserted that its absence was a 'fruitful source of ill-health', which caused 'the wheels of life...to clog, no organ does its duty properly, and if the seeds of disease are sown or breathed into a body weakened for want of exercise, it will find plenty to feed upon'.[42] Drill and variations on Swedish gymnastics, which did not require costly equipment or apparatus, as well as dancing and games, offered some opportunities to working girls at school and via club activities. And, as it became cheaper and more accessible, cycling would fulfil a mass demand, allowing increased access for girls to participate in outdoor sport.[43] Girls who could not afford to attend courses and classes were also encouraged to follow the less appealing regime of 'kitchen gymnastics' and to 'turn their attention to housework as a means of physical exercise'; 'Many a peevish, discontented, sallow young woman would be transformed by a good, liberal dose of housework.'[44] 'Daily toil', declared Anne Mahon in the *GOP*, 'is the most beneficial exercise in the world, if rightly performed, and will do more to keep you in trim than any other kind of occupation – except it be gardening or working in the open air'.[45]

Girls were advised to show restraint and care when taking up sports and exercise regimes. If working girls were warned less about this it was likely to be because it was deemed less relevant to them due to their limited access to such activities, rather than because they were considered fitter or stronger. Physicians warned consistently about the dangers of overstrain and the theme of 'moderation' ran like a red thread through debates and advice literature of the period with respect to both boys and girls, though for girls it remained entangled with concerns about their future roles as mothers and the risks of making them masculine.[46] Lowe, for example, advised young women to adopt a gradual approach, 'judicious exercise' and the need to develop 'general fitness' before embarking on a specific form of sport or exercise regime.[47] In her preface to Emma Walker's *Beauty Through Hygiene*, Kenealy advised discretion in the amount of time devoted to exercise, and the importance of achieving a balance between muscular activity, brain work and home and professional activities, and of stopping short of fatigue: 'By habit and practice she may be able to convert herself into a mere muscular machine, but she can only do this at the expense of health and looks and qualities.'[48] Mary Whitley cautioned girls in their sporting exploits to

> avoid the grotesque behaviour of the inconsistent creature, who affects a fine contempt for man, and yet attires herself in a close imitation of his garments, crops her hair close, and rides a bicycle in so-called 'rational' dress. I would encourage in all my readers a spirit of sensible self-reliance and independence, and would say to them, lead healthy, out-door, open-air lives by all means, but – and this is a very grave 'but' nowadays – be sure you do not run recklessly into danger, not injure your health seriously by the violence of your physical exercises.[49]

Ada Ballin's introduction to *Sports for Girls*, published in 1900, outlined the various sports available to young women in a more upbeat manner. Ballin, writer on health and editor of the health and beauty sections of the *Lady's Pictorial*, founded the monthly *Womanhood* in December 1898 aimed at the 'New Woman' of the 1890s, in response to what she perceived to be a gap in catering for the intellectual and physical needs of educated women.[50] While urging 'moderation in all things', Ballin aimed to foster an interest in a wide range of sports amongst girls, 'as tending not only to improve the condition of the girls themselves, but also, by strengthening them, to improve the condition of future generations'.[51]

Many features contained advice not only on the suitability of specific sports, but also on practice and technique, explaining how to build up activity steadily and carefully. Competitive sports tended to be discouraged for girls, though daring and exhilaration were often encouraged. Sir Benjamin Ward Richardson, a pioneer of the sanitary movement and an early promoter of cycling, was agnostic in advocating recreations that enabled the body to be developed to the readers of the *GOP*, so long as they did not 'vulgarise', impinge on domestic duties or involve competitiveness. With regard to his concerns about national regeneration, however, he cautioned that 'if the race is to progress they must some day become mothers, that they must undertake special maternal duties'.[52] In Richardson's view, walking was the best exercise, though he acknowledged that it was monotonous, and swimming, lawn tennis, rowing, cricket, golf, riding, hockey, archery, croquet, skating and rinking were deemed beneficial within limits. Horse riding, he regretted, was restricted to better-off girls, but was a splendid form of recreation; 'I do not know that it is more healthful than cycling, but it is perhaps more exhilarating, and there is about it a touch of adventure which is by no means a contemptible part of all recreations'. Golf, 'somewhat wearisome' was more appropriate to 'women advanced in life'.[53]

The richly illustrated 500-page *Girl's Own Outdoor Book*, edited by Charles Peters in 1889, contained information on numerous sports, as well as holidays, travelling, gardening, rearing fowls, botany and bird watching, and recommended the outdoors unreservedly.[54] While horse riding was to be conducted side-saddle on the grounds of decorum, the horse in question was expected to be lively and the girl to be very much in control. 'Ask Medicus whether a smart canter over the downs, or a swinging trot along some country lane, will not bring a set of muscles into play that often otherwise remain unused, rouse the torpid liver, and plant many a pale cheek with roses.'[55] While 'Medicus', aka Gordon Stables, supported a smart canter and cycling, which he described as a cornerstone of health, he suggested that in general exercise should be regular and moderate and urged caution when taking up gymnastics, 'and could point to cases of girls who had been injured constitutionally and for life, by what I may term this folly of rushing the gymnastic cure for debility... Don't let me catch you going and joining a gymnasium class... without having undergone a preliminary training.'[56] As Sally Mitchell has pointed out, magazines such as the *GOP* tended to preserve gender norms, despite their advocacy of an enormous range of sports. The *GOP*'s illustrations of tennis matches or girls' cricket often resembled garden parties rather than demonstrating vigorous activity, with girls dressed in long, flowing skirts and flowered hats, dainty and decorous, and remarkably stationary. An image of a girl cycling in the winter, skates on her arm, depicted a perfectly poised and exquisitely attired, fashionable young woman, lacking in any form of dynamism (see Figure 3.1).

By 1909 Amy Barnard's *The Girl's Encyclopaedia* covered a vast range of sporting activities, including many team sports such as hockey, basketball and cricket. Vigorous housework was also recommended and elocution and singing counted too, as these exercised the lungs and developed breathing capacity. In taking on these activities, Barnard counselled, however, that the girl must avoid overstrain and fatigue,

"BICYCLING TO HEALTH AND FORTUNE."

Figure 3.1 Lawrence Liston, 'Bicycling to Health and Fortune', Part II, 'The Rider', *Girl's Own Paper*, XIX: 939 (25 December 1897), facing p. 198 (The British Library).

follow a system which suited her age, physical powers and circumstances, dress suitably, and practice regularly, aiming to do a little activity well and frequently.[57] Barnard felt able to report that whereas only a few years ago there was no scientific system of physical culture for girls, 'Now every school of repute has its gymnastic classes, its drill hall, and playing field; indeed ill-health is regarded as almost a disgrace to a girl. Frequent unfitness is suspected of being allied with idleness, greediness, or some indiscretion in diet or management of self.'[58]

By this time physical culture was not only increasing in popularity but was also becoming more institutionalised, generating periodicals, books and gymnasia; it was promoted most notably in Britain by Eugen Sandow, who created a body-building empire, including a gymnasium catering for ladies.[59] Advice books increasingly included lengthy sections on exercises for girls, richly illustrated with photographs of individual movements, and physical culture became a popular topic in magazines catering for young women.[60] The *Girl's Realm*, notably its regular features writer on exercise, the feminist novelist and playwright Emily Morse Symonds, promoted a dynamic approach to exercise, and encouraged physical culture for girls unreservedly alongside its advocacy of other sports and sporting prowess more generally. Early issues contained articles on bicycling, fencing, golf, gymnastics, swimming, riding, tennis, mountaineering, skating and horse riding, as well as instalments of two series 'Girls who Excel at Sports' and 'True Stories of Girl Heroines', while its *Annual* for 1900 included features on ice hockey for girls, cycling in the Alps and hunting.[61] 'Are you one of the few girls of the present day who has never been introduced to the manifold joys of a gymnasium?', Symonds inquired. 'If so, it is high time that your experience should be enlarged, and the defect in your education remedied.'[62] The magazine went on to explain the structure of a typical class commencing with musical drill, followed by gymnastics, working with parallel bars, the vaulting horse, rope climbing and so on, concluding with the giant's stride. The text was accompanied by a set of illustrations, showing girls engaged in challenging exercises, active and vibrant, in sharp contrast to the static images presented by the *GOP* (see Figure 3.2).

> As time goes on, and the first awkwardness wears off, you will find that the fascination of gymnastics grows daily stronger. It is pleasant to feel the muscles strengthening, and the limbs growing lithe and supple, while the feats that once filled you with despair gradually come within the bounds of possibility, and then suddenly seem quite easy. And there are few things in life so entirely satisfying as the mastering of some physical difficulty ... Head-aches, 'nerves,' dyspepsia, bad circulation, and a host of other ailments, are frightened away by the vaulting and jumping, while the fever germ finds it extremely difficult to effect a lodgement anywhere about the person of an enthusiastic gymnast.[63]

It was argued by those eager to encourage girls to take up exercise and sport that this would benefit health, intellect and beauty as well as

THE PRINCIPAL MUSCLES USED
IN THIS EXERCISE ARE THE
ABDOMINAL GROIN AND
QUADRICEPS.—THIS IS ONE
OF THE SPECIAL EXERCISES
PRESCRIBED BY MR. SANDOW
FOR INDIGESTION AND
OBESITY.

Figure 3.2   E.M. Symonds, 'Physical Culture for Girls', *The Girl's Realm Annual* (November 1898–October 1899) (London: Hutchinson, 1899), p. 62 (The British Library).

having remedial and therapeutic benefits, despite varied opinions on what constituted appropriate activities and how far these were to be moderated. Increasingly – and in sharp contrast to Kenealy's Clara – exercise became associated with positive mental and personal attributes, as well as exuberance and pure enjoyment of physical activity, strength and agility. While moderation and careful application were insisted upon, the literature also emphasised the thrill of learning and improving, of building control and understanding bodily performance and movement. As early as 1879, J. Hamilton Fletcher had argued that proper training would 'render girls strong, self-reliant, able and skilful, then we should find that the physical cowardice of women had vanished, that their moral courage was strengthened, their minds rendered enterprising, cool, and liberal'.[64] The *Girl's Realm* aligned working hard to excel at games with working academically to excel at geography or arithmetic, urging girls to become strong and active, by conducting exercises which, while not difficult, brought the will to bear on them, and required vigour and thoroughness.[65] In addition to making muscles strong and responsive, the Stenhouses stressed that games would 'help us to judge how far one thing is from another,... and they should teach us to be good-tempered and patient and fair with each other – to win without boasting and to lose without being angry'.[66] Drill, meanwhile, would ensure that no part of the body would be unduly strained and improve concentration. Exercise was not only good for physical development, declared Mary Humphreys, but also had 'far higher influences... It is good for your mind and for your moral nature generally. Your powers of endurance and determination are strengthened, and a full mind-control of the body is established.' It promoted aesthetic sense, harmony and a love of beauty.[67] Amy Barnard also argued for the multiple benefits of exercise:

> While attaining health, the methods of physical culture... not only discipline the body, but develop character. Proper exercise encourages self-control, and reacts beneficially on the mind... as the mind gains perfect mastery over muscular movements, fearlessness, self-reliance, promptness, dexterity, accuracy of vision, touch, and balance are cultivated.[68]

American journalist and author, Christine Terhune Herrick, suggested in 1902 that athletes built a wide range of skills, sport promoted a sense of 'physical consciousness' and awareness of the need for proper diet and rest. She also disputed the argument that athleticism led to 'the decline in manners of the girl of the period'. Rather it encouraged 'respect for others, accuracy, self-control, patience, conscientiousness, honour' and, perhaps most significantly, 'moderation'.[69]

Cycling for health

Cycling offers an excellent case study of the debates involving exercise for girls. Not only do we have numerous printed articles, books and pamphlets on the topic to draw on – it was an apparently inexhaustible source of copy particularly in the mid-1890s – but debates on cycling were emblematic of wider concerns on the purported medical drawbacks and proposed health benefits that figured more generally in debates on exercise and sport for women and girls.[70] There was intense discussion, compressed into a relatively short time frame, in medical and lay media, including cycle periodicals, on the pluses and minuses of cycling, which was configured as being a potential vehicle, literally as well as figuratively, to propel women forward in their pursuit of independence, mobility and enriched recreational activity as well as physical and mental good health. The fact that cycling took place primarily on roads and in parks placed female cycling firmly into the public domain – it was a highly visible activity and, linked intrinsically to the question of appropriate clothing, triggered debate in particular about the appropriateness and utility of rational dress.[71] David Rubinstein has suggested that the bicycle gave women, particularly the young and unmarried, three benefits: the process of participating in active recreation received a considerable boost in the 1890s as cycling increased in popularity, it offered a 'hearty blow' to the system of chaperonage, and, most significantly, it defeated 'conservative opinion symbolised by a woman riding a bicycle'.[72] Additionally, as it spurred growing acceptance of the fact that outdoor exercise was good for the health of women and girls, it became embedded in wider cultures and practices of health and signalled changing ideals of the female body, which distanced itself – as did Clara – from the ideologies and deportment associated with Victorian femininity. Girls' magazines and advice literature on health and recreation – as well as annuals and stories – engaged enthusiastically with the craze, exposing girls to the huge potential of this new sporting opportunity and its advantages for health.

In July 1895 the *Queen* magazine declared – as the cycle craze was about to peak – that cycling stood out amongst other sports as a means of improving women's health. It could be done all year round, the only equipment required was a bicycle, women could cycle alone, and it encouraged 'independence of the cycle', pluck, nerve and self-reliance. 'To women who, from choice or press of circumstances, have gained independence of work in their life, the use of the cycle gives independence and freedom in their play.' Instead of returning from the office or workshop by omnibus or rail straight to household duties, the cycle offered the opportunity for exercise and recreation. 'Thus, by increasing the health of both body and mind, and by the training of nerve and the higher moral qualities, the use of the cycle becomes a more important factor in the great movement characteristic of this century – a movement of which we see, as yet, merely the beginning.'[73] A few years later *Queen* magazine, demonstrating the health achievements of cycling, praised the sport which had removed hundreds of women from their sofas 'sighing and groaning over their health, [who] are today pedalling their twenty miles and feeling all the better for it'.[74] As Clara's story exemplified, however, the woman cyclist was not without her critics. Writer Eliza Lynn Linton, who for several decades had been castigating girls for their declining standards of behaviour, in large part due to their sporting endeavours, described in the *Lady's Realm* in December 1896, how the woman cyclist had lost the 'sweet spirit of allurement' and became intoxicated with 'unfettered liberty' and the 'modern passion for imitation of the manly life'.[75]

By the mid-1890s there had been a huge boom in cycling, with an estimated 1.5 million cyclists in Britain, as the bicycle took the lead in triggering consumer interest in 'novel items'.[76] Cycling periodicals were hugely popular, including several catering for female cyclists, and newspapers, society journals, general interest and girls and women's magazines all contained cycle news, and a vast range of guidebooks for cycle touring was published. By 1898 it was estimated that there were over 2,000 cycle clubs in the UK.[77] Growing demand for ladies' bikes indicated the sport's increased popularity amongst women and girls. 'Ten years ago', an article in the *BMJ* declared, 'a woman on a tricycle was a *rara avis* to be hooted at by small boys in the street; to-day the manufacturers cannot cope with the demands for ladies' bicycles'.[78] As the ordinary bicycle or Penny Farthing and the tricycle, was superseded by the Safety Bicycle, with two equally sized wheels and pneumatic tyres, bicycles became safer and more comfortable and increasingly popular with women and girls. One-third of all bicycles ordered in Britain in 1896 were ladies' cycles, up from a mere 50 in 1893.[79] Constance Everett-Green, a regular contributor to cycle magazines, ascribed the popularity of the bicycle and its leap into the forefront of fashion in the spring of 1895 to the 'more fashionable and aristocratic of our sisters', at a time when the propriety of women riding bicycles was still questioned.[80] By 1895 the bicycle was, according to the *GOP*, 'unquestionably quite the rage', and for girls likely to be 'a source of pleasure beyond anything she has already had'.[81] Mrs Humphrey noted in the illustrated monthly, *The Idler*, in the same year that 'Hundreds of gently nurtured girls' were seen cycling in Battersea Park 'some of them expert enough, others still in their novitiate, and many of them accompanied by mothers who have had perforce to take to cycling in order to perform their duties as chaperons'.[82] 'Cycling', declared one writer in the *GOP*, 'is practically the only pastime by which it is possible to make public the appreciation of athleticism'.[83] By the late 1890s cycling amongst ladies of rank had declined somewhat, as they turned towards the increasingly fashionable motor car, but by then it had become a popular pastime for upper- and middle-class women, and toward the turn of the century, as the cost of cycling fell, it was taken up by working-class girls.[84] Everett-Green declared that by 1898 the absolute fashion for cycling was on the wane, but by then 'The country at large acknowledges cycling as a usual and suitable exercise for ladies.'[85]

As cycling increased in popularity, this triggered a vigorous exchange about its dangers and potential benefits in medical journals, the cycling press and lay periodicals.[86] It was rare for physicians to wholeheartedly deplore or praise cycling, though Dr Edward Beardon Turner, Chairman of the General Committee of the National Cyclists Union and author of a ten-part series published in the *BMJ* on cycling, health and disease, referred to many of the conditions linked to cycling – as well as the 'bicycle face' – as 'myths'.[87] Concerns focused initially on the risks of 'furious riding', accidents and collisions, before shifting to the question of whether cycling inflicted upon both men and women more durable complaints and chronic ailments related to posture, pressure, the over-stimulation of vital organs, overtaxing particular muscles or overexertion. Cycling was described as an 'epidemic' with its own set of diseases and conditions – some peculiar to men and many shared between the sexes – including 'bicycle hand', 'cyclists spine', and vibration and fatigue fever. The *BMJ* fielded numerous queries on cycling's association with hernias, varicose veins, haemorrhoids, urethral stricture, and cardiac and nervous disease.

Women and girls, however, were the subjects of particular concern, the female perineum, being, as Whorton has put it, 'the object of such extraordinary medical solicitude throughout the Victorian era'.[88] The cycle craze in Britain trailed that of America and France, as did the concerns about its effects on health. The New York obstetrician, Robert

Dickinson, concluded that cycling might offer 'special rewards' aside from general exercise, leading, as much of the work of cycling was done by the legs, to improved blood flow and tone in the pelvis, which would relieve a range of female complaints – chronic pelvic disorders, varicose veins and hypersensitive local nerves.[89] However, he warned, 'It is suggested that the friction of the saddle may lead to sexual excitement; and, although the author is only aware of one instance, still he thinks that a proper arrangement of the saddle should be provided to prevent any possibility of contact with the clitoris.'[90] In the same year, Dr A.T. Schofield explained to the readers of the *GOP* that he had been consulted by young ladies with their mothers 'to whom I felt this new exercise would prove a real boon, if used in moderation; I dared not propose it, as at the time it was hardly supposed to be within the bounds of propriety'. In rather more delicate terms than Dickinson, he objected to the hard leather saddle 'that is very much like the section of a long-necked pear', advocating instead a 'nice padded cushion, flat in front, and perhaps rounded behind' but 'remembering also that the time of life when girls ride most is that in which their nervous system is still being evolved – that the invention of the pneumatic type is of the greatest value in lessening the constant vibration'.[91]

The question of the readiness of girls to cycle attracted particular attention, and some charged the cycle with 'leading young and innocent girls into ruin and disgrace'.[92] Robert Ingle, FRCS, warned in a letter to the *Cyclists' Gazette* in July 1895 that girls should not take up cycling before the age of 18: many were unfit due to 'inherited general or local weakness, expressed in imperfect bone formation, defective joint structures, or blood poverty, and others have unfitness induced and maintained by unfortunate conditions in their occupation, residence or diet'. The parts 'enclosed by the hip bone' were most likely to suffer, 'any interference with which may prove to be disastrous in future years'.[93] F.L. Gerald, MD, described how the bicycle craze, which was 'running at full blast', resulted not only in desecration of the Sabbath but also brought on irritation of the bladder and organs within the pelvis. In a frequently used analogy to the repetitive muscular demands of the sewing machine, which was also linked to the over-stimulation of 'carnal instincts', he predicted that it would cause a 'large amount of suffering with young women, unless moderation is used'.[94] Though he praised cycling as a form of outdoor exercise, A. Shadwell, inventor of 'bicycle face', pointed out that girls were particularly vulnerable to harm, as they 'set exceptional store by amusement, and they are, by nature, shy of saying anything about their health'.[95] Dr Frances Hoggan, Medical

Inspector at the North London Collegiate School for Girls, concluded that delicate women and girls in particular derived great benefit from cycling, but that any women 'suffering from any of the complaints incidental to their sex' ought to seek out medical advice before they took to cycling, a caution supported by many other medical authors, who also agreed that racing, competitions and strenuous training was bad for women and girls.[96]

While Edward Turner suggested in 1896 that 'There is no reason whatever why any sound woman should not ride either a bicycle or a tricycle', he went on to qualify this with the advice that women should not ride during menstruation or pregnancy nor for three months after confinement.[97] In the case of girls, objections had been raised to the use of the cycle on the grounds that it could cause enlargement and hardening of the muscles of the pelvic inlet, thus making parturition more difficult, but experience, Turner argued, had shown this not to be the case.[98] He believed that cycling would not cause 'a mechanical obstacle to the due performance of that all-important function', but also warned that a woman who had 'overtaxed her strength and squandered her reserve forces in the preparation for, and participation in, athletic competitions, for which by nature she is unsuited, is not one of those who would best be fitted to pass through the ordeal of motherhood with advantage to herself and her offspring'.[99] In 1896, reflecting the ambivalence of many doctors, the *Lancet* advised that the best approach was to wait for more information on the medical impact of cycling, taking into account the 'opposite opinions' on the effect of cycling on the pelvis and perineum, and pointing out that in some disorders – hysteria, neurasthenia, anaemia, chlorosis, dyspepsia, constipation and general cases of invalidity – 'the bicycle was likely to be advantageous'. However, 'In the meantime it would seem desirable that young girls especially, whose pelves are not firmly ossified, should either be advised not to use the bicycle at all, or at any rate to use it to a very moderate extent.'[100]

Concern about the impact of cycling on the organs of reproduction was relatively short-lived – despite the initial flurry of articles produced on this topic – and within a few years of women taking up cycling in any number most medical commentators agreed that little risk in terms of women's reproductive capacity was attached to this activity and that there were many potential benefits to health. The Countess of Malmesbury noted in 1898 that

With a few exceptions the medical profession adopted an attitude of extreme caution with regard to cycling for women, regarding it with distrust, and discouraging it in every possible way; but now there is a distinct reaction in its favour, and only a few of the obstinately blind are found in opposition.[101]

'The advantages of cycling are so great', argued one correspondent to the *Lancet* in July 1897, 'and none the less to women who contemplate marriage as well as those who may be living in their first years of married life'. Indeed 'Cycling has saved many lives in both sexes', and was 'one of the most enjoyable, popular and healthful exercises ever made fashionable'.[102] Another supporter suggested that cycling was 'the most perfect means of physical exercise' for women: 'The bicycle, affording a means of exercise in the open air, bringing into play almost all the muscles of the body while requiring but a gentle effort, deserves the attention of every hygienist desiring to obtain for women a suitable mode of muscular exercise.' Although cycling was also recommended for working women as their access to this form of exercise improved, it was suggested that it was particularly valuable for women who did not work and who therefore had little opportunity for the expenditure of muscular effort.[103] Women derived enormous benefit, one of its main proponents argued, from an exercise that 'increases the circulation, strengthens the muscles, develops the chest, relieves the lower limbs from the dragging weight of the body, and enables us to obtain far more fresh air than our powers would otherwise allow'.[104] Cycling, another enthusiast declared, was 'agreeable and fascinating', helped develop the muscles of the back and legs, and was 'one of the most health-promoting' exercises that women could undertake.[105]

Presenting at a conference on domestic hygiene on the theme of cycling for ladies, Edward Turner went further – inspired perhaps by his all-female audience – arguing that it was liable to improve 'not only the individual, but the race generally', and suggesting that there was an almost unanimous consensus of opinion among those 'best qualified to judge' that the health of women who cycled had improved, much more so than through engagement with other forms of sporting activity.[106] In general, Turner argued, healthful sports had seen women grow in stature, and seen the evolution of the 'Tall Girl', with strong expectations of race improvement, as occurred in the 'sons of athletic revival' 30 years ago and their offspring. 'To the athletic father and tall mother we must look for the perpetuation of a race which will rise superior to the Neurasthenia which is the bane of nineteenth-century existence, and teach the "Decadents" that after all the sound mind in the sound body is the better part.'[107] Women should ride, an article in the *Hub* proclaimed, because the 'bicycle is the best known developer of physical beauty in line and curve in woman, straightening bent shoulders and tired backs, raising the chest, and scattering nervous headaches to the winds'.[108] Henry Robert Heather Bigg, FRCS, an authority on spinal curvature, wrote how there had been a remarkable increase in the stature, development and size of limb of young women compared with their mothers, which he attributed in part to their taking up of a range of sports, notably cycling, which exercised muscles without fatigue, and stepped up circulation, respiration and perspiration.[109]

Dr William Hugh Fenton's influential article on cycling, which appeared in *The Nineteenth Century* in 1896, suggested that cycling might enable the levelling out of differences between the sexes in terms of physical endurance, allowing women to throw off some of their encumbrances, such as dress that handicapped movement, and to compensate for the lack of early training and muscles that had been disused for generations. 'Women are capable of great physical improvement where the opportunity exists.'[110] Fenton, a Harley Street physician and specialist in obstetrics and gynaecology, said little on this topic with regard to cycling. Rather, he focused on the ways in which women were slowly overcoming prejudice by means of their engagement in physical activities, suggesting that cycling would accomplish a form of 'revolution' in their muscular tone and powers. To achieve this, he added, they needed to approach cycling gradually to build condition and endurance. Most women, he concluded, had in any case 'set their faces' against the 'physiological crimes' of racing and record breaking.[111] In 1895 *The Review of Reviews* declared:

> There is no reason to think that a healthy woman can be injured by using the wheel, *provided she does not over-exert herself by riding too long a time, or too fast, or up too steep hills; and provided she does not ride when common sense and physiology alike forbid any needless exertion*... and there *is* reason, not merely to *think*, but to *know*, that many women are greatly benefited by the exercise.[112]

Lady Jeune, society hostess and journalist, also advocated cycling as a healthy pastime: 'There is no necessity for women to race, to go at an unduly rapid speed, or to ride their machines uphill; and if they forego these forbidden pleasures, they may bicycle in safety to the end of their lives. The modern machine is so light and well designed that there is no fatigue in riding it on a good road.'[113] Sir Benjamin Ward

Richardson suggested that 25 miles was a good day's ride for an accomplished rider, and urged men and women to dismount and walk when climbing long and steep hills, which brought new muscles into play and saved strain on the muscles of respiration.[114] Girls, however, Richardson argued, should only begin to cycle regularly at the age of 17, and only if 'they are strong and well formed. In training, and ever afterwards, they should be taught to sit straight up ... and always to have the dress perfectly free around the waist and chest.'[115] Fenton, meanwhile, urged that the young growing girl be watched and warned that 'her youthful keenness should not carry her beyond her powers of endurance and easy recuperation'.[116] The *BMJ* confirmed this view, stating that moderation needed to be practised by all, young or old, male or female, 'but should more particularly be impressed on young girls. Most boys have all their life been accustomed to hard exercise, and so undertake a new sport more or less prepared; but the reverse is often the case with their sisters.'[117]

Health advice columnist to the *GOP*, Dr Gordon Stables, counted himself as a pioneer of cycling, which he regarded as an excellent form of exercise for girls.[118] The health benefits were manifold; cycling tackled rheumatism (his own, he noted, had been cured by cycling), ennui and low spirits, debility of the blood, constipation and indigestion, overweight and sleeplessness. It was also recommended as an aid in curing nervousness, as part of a regime of diet, cold bathing, brisk exercise and positive thinking that Stables was particularly fond of. He was more cautious in suggesting that girls should not learn to cycle under the age of 15 as they were still growing, though he urged weak girls to aim to use a tricycle 'if the muscles of the limbs are tolerably firm; if the face be not preternaturally pale' and if they had good digestion, and could run up steep stairs without discomfort; 'having commenced this pleasant form of exercise, you are likely to get stronger and healthier every week'.[119] He instructed girls to wear light corsets to avoid pressure on the vital organs, reserving 'hornet-waists for the promenade, the afternoon tea, or concert-room but, if they value health, let them be more in accordance with nature while riding'.[120] He also suggested that girls were more likely than boys to attempt to do too much when taking up cycling. In addition to lauding the health benefits of cycling, the *GOP* advised in numerous articles on how to dress for cycling, cycle maintenance, where and how to cycle, cycle clubs, fancy cycling and cycle gymkhanas. Benjamin Ward Richardson advocated cycling as a counter to urban life and as a means of halting national degeneration, but reminded young women that the dangers of passing 'in recreation beyond a certain bound of natural womanly duties, is to pass into a sphere with which such duties are utterly incompatible'. Cycling, he concluded, was a good form of exercise for women provided that they did not overexert themselves or enter competitions.[121] Taking up the theme of moderation, N.G. Bacon, Honorary Secretary of the Mowbray House Cycling Association, concluded that

> Provided the cycle is used as a health-preserver, and not as a health-destroyer, the nerves are quietened, the energies are awakened and quickened, the muscles rested, the circulation is stimulated, and the feeling of physical satisfaction which pervades the system insures calm and invigorating sleep. Within moderation, i.e., when the powers are never over-exerted, but properly exercised, cycling strengthens and invigorates the system.

Overexertion, he went on, led to loss of appetite, restlessness and sleeplessness. It was also important to ensure that the saddle was at the right height, that the handlebars were properly positioned, the pedals were in reach and to avoid overheating when cycling.[122] Dr Lawrence Liston, another of the *GOP*'s regular contributors, argued that 'it can only exceptionally be the case that a girl of ordinary physical endowment cannot ride a bicycle', which would lead to 'the feeling of well-being and increased physical strength', though he also noted that care and discretion needed to be used to avoid overexertion and, like Stables, recommended that cycling should be prohibited to girls under the age of 14 or 15. Girls were advised to cycle regularly rather than to make sudden calls on their resources.[123] The desired end of health, increased appetite, rosy cheeks, and increased physical strength, was likely to be reversed, Liston warned, 'through ill-directed over-exertion'.[124]

In 1889 the *BMJ* carried a leading article promoting cycling and health, excluding the 'silly craze' of racing and record breaking, arguing that 'Cycling as a therapeutic agent has a considerable future.'[125] And while cycling continued to have its detractors – one cautioned about a young lady who 'cycled so much that the constant vibration at last unhinged her nerves completely...' and another that 'guiding and controlling a machine is something of a nerve strain' – most commentators agreed by the late 1890s that it was beneficial in boosting general health, mobility and fitness and had distinct therapeutic advantages.[126] 'The bicycle', one anonymous author asserted, 'is to relieve the female sex, who used to be all *nerves*, of their nervousness. It is to add to their self-reliance, a quality whereof the *new* young woman has already amassed a goodly stock.'[127] Turner urged the need for cyclists to be in sound health to start with; in specific ailments 'cycling, properly regulated, acts like a charm in restoring health; there are others in which to mount a bicycle would be simple suicide'.[128] He argued that it would lead to a 'favourable modification' in women's physical condition – encouraging them to train, to modify their meals, to improve their habits of attention and gain in courage and self-control.[129] Turner recommended cycling for functional diseases and those arising from insufficient exercise, as well as for ailments such as gout, rheumatism and indigestion. 'In the bloodlessness of young girls it sometimes does more good than pints of iron drops, though in such cases moderation is most essential until the heart is well drilled in its new work, and very few instances of pure "nervousness" survive a regular course of bicycle rides.'[130] Fenton argued that cycling was an excellent antidote to anaemia amongst adolescent girls with its associated 'languor, morbid fancies and appetites'; it oxygenated the blood, improved circulation and got the patient out into the air and sunshine. Improved circulation was particularly beneficial for shop girls, who were obliged to stand for many hours: 'the aches and pains which would have shortly made an old woman of her have gone, and a sense of exhilaration and relief has taken their place'.[131] Hysteria, sick headaches, anaemia and neuralgia, 'as well as imaginary ills' were benefited by cycling, while for 'girls earning their own living and confined all the week in places of business, certain cycling clubs instituted on their behalf are a great boon'.[132]

Working-class girls were urged to take up cycling 'for health, for convenience, and for freedom', and a number of schemes were set up to promote cycling amongst those unable to purchase a bicycle.[133] Working girls' magazines, such as the halfpenny *Girls' Best Friend*, promoted cycling and offered hints on choosing bicycles.[134] Some employers provided bikes for the use of shop girls so after work they could cycle in parks or the countryside, and in the East End of London clubs were set up to loan bikes, 'a real boon to women-workers toiling long hours in the close, used up atmosphere of London shops'.[135] 'To the working woman, confined all day in an office or store, it has already proved to be a friend in need.'[136] Second-hand bicycles were cheap, and by the mid-1890s working girls were reported to be cycling in London in their free time. In May 1893 the Mowbray House Cycling Association was founded, with the aim of supplying cooperative cycles to members, by W.T. Stead, editor of *The Review of Reviews* and promoter of equal opportunities for women.[137] Its membership was not uniformly supportive of working-class cyclists, however, and a 'Rational Rational-Dresser' who,

while feeling for her sisters of small income who cycled, advised them
to learn

> that their cobbled and bungled nether garments are simply ludicrous,
> and if they cannot save up the modest sum necessary to purchase
> prettier garments that are the work of professional hands, I would
> advise them to accept the skirt, short if you will, as another of the
> many drawbacks of poverty.[138]

Snippy comments like this aside, Mowbray House Cycling Association,
including its President, Lady Henry Somerset, declared cycling to be a
route to emancipation for women, 'in the new freedom granted her by
public opinion to share the harmless, healthful, open air amusements
of her brother'. A revolution was also being wrought for 'the stooping,
narrow chested, anaemic looking girls who crowd the offices, factories
and shops of all our cities', for whom the countryside would soon be at
their doors by means of 'that swift steed of steel, the bicycle'.[139] The *GOP*
advocated the bicycle for a range of working women – teachers, district
nurses, seamstresses, girl clerks and shop workers 'are all the better fitted to be bread-winners by their brief but happy glimpse of the country'
which cycling facilitated.[140] The periodical, *At the Sign of the Butterfly*,
the mouthpiece of the Mowbray House Cycling Association, looked forward to the year 2000 being marked by the dominance of 'all-round and
well proportioned' women, leading a 'full, free, untrammelled physical
life to which her economic independence opened the way'. This would
contrast with the situation in the nineteenth century, when

> The complicated system of bondage under which the women were
> held perverted mind and body alike, till it was a wonder if there
> were any health left in them...Up to fifteen she might share with
> her brother a few of his insipid sports, but with the beginnings of
> womanhood came the end of all participation in active physical
> out-door life.[141]

The idea that women and girls had limited funds of energy to be
apportioned to mental and physical activity was alluded to in advice
on moderation, regular practice and correct approaches to cycling, and
instructions to avoid overexertion and scorching, cycling for too long
or too fast, which could lead to exhaustion, and loss of sleep and
appetite, defeating the object of the exercise of cycling. But it was dealt
with pragmatically as part and parcel of instructing girls in 'method',

involving instruction in cycling techniques and imparting the idea that exercise needed to be built up steadily. 'A discreet rider should never allow herself to be taxed to the limit of her powers but should always have some reserve force in hand.'[142] Though A.T. Schofield had some reservations on the suitability of cycling for women, he argued that as a balancing form of exercise, 'it develops capacity and strengthens the brain, without any risk of strain . . . it is no small recommendation of this new exercise that it is indubitably not only a healthy means of relieving the brain of over-pressure, but that it also exercises and strengthens this organ at the same time'.[143] Cycling was also praised – because of the imperative to stay balanced – for engaging all muscle groups. While situating this largely in the realm of work, Anson Rabinbach has considered the ways in which theories of thermodynamics were applied increasingly to physiology and then to ideas of energy conservation by the 1880s with regard to the 'human machine' and 'muscular force' and by the 1890s in relation to the problem of fatigue.[144] Yet discussions on cycling engaged to only a limited extent with such concepts, suggesting that fatigue could be avoided by proper practice. Indeed, W.H. Fenton referred to the value of cycling for women in 'working off their superfluous, nervous and organic energy', rather than causing exhaustion.[145]

Women who wished to take up cycling were advised to do so in consultation with a doctor who would lay out a plan to build up strength and endurance. Cycling for women thus became an opportunity for doctors to lead a movement away from the obsession with female frailty, nervous complaints and hysteria to manage a new process of achieving health. The danger, as reported in *The Speaker* in May 1896, was that this would in the end rob doctors of custom.

> It is conceivable that the next generation but one may give us an average type of womanhood wholly free from the nervous ailments which make so many women to-day a prey to morbid emotions. There will be a good deal less doctoring in that happy time . . . which may deprive Harley Street and Wimpole Street of consulting rooms in the year of grace 1950.[146]

Women's take up of cycling was also conceived of as an opportunity to promote hygienic practices, a spur to establishing new regimes and approaches to health. Importance was attached to the wearing of a particular kind of garment made from wool, flannel or tricot, and to bathing following exercise:

At all events, on returning home the fair rider should follow the
example of all men who are accustomed to exercise: she should
change her moist underwear at once, and adopt, in entering on an
unfamiliar form of exercise, the hygienic habits without which util-
isation of the muscular powers, so beneficial to health, may be a
source of immediate danger.[147]

While rational dress was never widely taken up in Britain and dress
adapted slowly to cycling, women were advised to wear shorter skirts
with loser bodices and to discard petticoats and long, constricting or
close-laced corsets.[148] 'Lady Doctor', Miss Crosfield, stressed moderation,
'disuse of the corset, and recommends wide shoes, an upright position, and woollen underwear', a diet of meat and vegetables, and urged
riders not to cycle on either an empty stomach or immediately after
a meal.[149]

Ada Ballin provided a detailed breakdown of clothing, diet and skin
care for girls taking up sports of all kinds, and urged young women 'to
have the courage to say when they are tired, and not strive to emu-
late feats which are easily accomplished by their stronger male friends'.
In the case of cycling, they should build up gradually and in two to
three months they would 'be able to do twenty or thirty miles a day
with the greatest ease'.[150] This was fairly typical of the distances suggested for experienced girl cyclists, which vastly increased their mobility
and took them to new scenes and places.[151] The diary of a girl recording
her cycling exploits in and around Yorkshire in the mid-1890s described
how she participated in cycling holidays involving train travel and
extensive tours of the Peak District, in all kinds of weather, sometimes
with companions, but often alone. She relished cycling and the scenery
no matter how challenging the conditions, brushing off a number of
minor accidents and spills and rejoicing in her new cycling outfits.
In February 1895 she set off by herself, with snow banks on the side
of the road, and a rough north wind: 'had a lovely spin beautiful day,
sunshine, but very cold, road like a cattle track...' A few days later she
cycled 16 miles, 'Roads dreadful inches deep in mud & slush', had a
skating lesson on the river, and then cycled home, 'got splashed up
to my neck, but jolly time altogether'. In 1895 she cycled some 1,485
miles.[152] Bim Andrews, born in more straightened circumstances in
Cambridge, reported how in the mid-1920s, following evening classes
in shorthand and typing, she took up a job as a higher grade office
worker earning 17 shillings a week. Her first purchase was a new bike
for £5,

to be paid for on the never-never at 2s a week. It had down-swept racing handles, and we wore trousers for riding... It seems as though I was getting free from the need to conform, for these ugly garments did cause a commotion at home and in villages where we stopped to rest and look. They made sense to us, however, as cycling kit, and so we took over where Mrs Bloomer left off. We covered a good many roads in East Anglia on our Sunday trips.[153]

Through their letters pages, cycling journals empowered girls who put forward their own views and joined in lively debates on the benefits of cycling and cycle dress. One correspondent, a young woman who had tried skirts and knickerbockers for cycling, came down on the side of a narrow tailor-made skirt, short enough to be clear of the pedals; she abandoned knickerbockers because there was nothing to be gained in wearing them, 'except perhaps a certain notoriety among one's acquaintances, and the candid and unprejudiced opinion of the little boys on the street'.[154] Miss Lillias Campbell Davidson, President of the Lady Cyclists' Association, encapsulated the feeling of independence and freedom which cycling meant for young women, showing how health became bound up with a sense of space and free movement outside the home or other closed spaces inhabited by girls – the factory, school or more regimented clubs and societies.

> There are few pastimes for girls that can equal, much less surpass, cycling. It is not only one of the most delightful amusements in the world, it is one of the most-health giving. The girl who sees others skimming past her on wheels with a swift, smooth motion like the swoop of a bird down the wind, will be fascinated by the look of ease and lightness with which they flit away into the distance, while she stands to watch; and she will long to be on wheels, and flitting too.[155]

By the late nineteenth century, cycling, along with many other sports, had been taken up by large numbers of girls and, while in its early stages it was closely associated with the craze amongst upper-class women as well as the New Woman in popular literature and imagery, by the mid-1890s its popularity had spread to middle- and working-class girls. Debates on the relationship of health and cycling would continue – but the immediate panic which highlighted the risks of gynaecological disturbance, the overdevelopment of particular muscles, overstrain and scares about the production of girls with bicycle faces quickly abated, and apparently had little impact on girls' take up of cycling. As Kathleen

McCrone has argued, medical and social authorities quickly adapted to the *fait accompli* that marked the take up of cycling by women.[156] The Countess of Malmesbury declared cycling 'one of the greatest blessings given to modern women', associated as it was with the emergence of the healthy, outdoors girl and tied to predictions of improving health and strength for women. 'A necessary element in the life of most girls in the county, and a never-failing source of recreation to those who live in towns', cycling was declared 'a favourable influence on the health of those who indulge in it, with discretion', leading to improved appetite, the ability to sleep soundly, 'the reddening of poor pale cheeks'; 'the feeling of well-being and increased physical strength tell loudly the tale of restored and perfected health'.[157] Cycling was configured in the broadest sense as transformative – linking embodiment, pleasure and emancipation with physical and mental fitness – as it became a popular feminine pursuit, prefacing by several decades the sense of fulfilment and enjoyment Jill Matthews has associated with the Women's League of Health and Beauty.

> And what pleasures are, none but those who have tried them for themselves can really know. The feeling of active movement, of the power of free locomotion, the thrill of healthful exertion, and the bounding of the pulses as one speeds along some level stretch, or shoots swiftly down some steep incline... A new world of enjoyment is unlocked to the woman who finds herself a-wheel... The nervous invalid, the victim of neuralgia and sleeplessness, the sufferer from a thousand nameless ills, finds herself suddenly endowed with health and vigour hitherto undreamed of, and a vitality she has never known before. Countless new interests present themselves, and she learns, to her surprise, that cycling is a door that leads to many paths of pleasure.[158]

By aid of 'your wheel', N.G. Bacon suggested in the *GOP*, girls could improve their historical, geographical and botanical knowledge, but above all

> We are introduced to the grand and inspiring world of Nature; drink in her beauties and revel in the myriad joys she so generously showers upon us.... After a rapid whirl on our wheel, a little tired and dusty, perhaps heated and restless, we alight, forsaking our mount for the nonce, and flinging ourselves down on the grass, we yield to the sublime thoughts which ever overtake those who are receptive. A love

of Nature has the power to rid our brains of all cobwebs, whether they come from the schoolroom, the nursery, the drawing-room, the dining-room, or the domestic cares of the kitchen; all can be swept away if Nature-worship is innocently pursued.[159]

Although girls were cautioned to be moderate in their efforts and to avoid overexertion – and thus fatigue – this was part and parcel of the general approach to the take up of cycling and the acquisition of new skills, such as getting on and off the bike safely and balancing, and much of the advice made sense for women and girls who were likely to have had little prior experience of sport and exercise. Many supporters of cycling claimed that it improved health or that it had therapeutic benefits, particularly with regard to nervous complaints. Though medical debates on cycling engaged with the disorders and diseases of female adolescence outlined in Chapter 1, they did not dwell on them for long, and cycling was seen by many as a means of combating them. Cycling, one physician concluded,

> gets them out of doors, gives them a form of exercise adapted to their needs, neither too violent nor too passive; one very pleasant, one that they may enjoy in company with others or alone, and one that goes to the root of their nervous troubles; for we are beginning to realise that these do not for the most part have their origin in woman's particular anatomy and physiology.[160]

Conclusion

By the end of the nineteenth century it was regarded as an accomplished fact that the physical make-up of the modern girl had changed dramatically for the better, with much of the debate about this change taking place in accessible lay publications, including material directed at girls. More exercise and better nutrition made girls 'taller in stature, stronger in build, healthier in look, freer in gait, more self-assured in action' than the previous generation.[161] Many girls, it was noted in 1895, participated in a huge range of sports, spent time taking walking tours, sea bathing or mountain climbing, and

> indeed, they are not now excluded from any healthful athletic game in which their brothers take part. It is possible that there is some danger that girls will overdo all this; but, on the whole, the sex has

gained immensely from the change of thought in regard to out-of-door games and exercises.[162]

Although cycling and other sports often involved team work or participation in clubs or group activities, exercise was presented in many ways as reliant on individual drive and application in the pursuit of personal health, and a reflection of good character. As Sally Mitchell has also pointed out, the 'atmosphere' regarding girls changed significantly in the 25 years separating two editions of Marianne Farningham's popular advice manual *Girlhood*. The 1869 version argued that girls must acquire 'womanliness' and 'dignity' in their late teens, 'the hoydenism, the frolic, and the exuberant mirth will now become unseemly'. By 1895, the language had changed and a new section had been added:

> There are very few girls indeed who are kept back by the fear of being called 'tomboys'...they can become strong and vigorous and yet retain that essential womanliness of thought and manner which is their greatest charm. So they take their morning bath, and sleep with their windows open; they eat plain food...dress comfortably, and ride on the outside of omnibuses and trams...And when newspapers call attention to the fact that young women are taller, and young men shorter, than ever, they smile, and are not surprised.[163]

'There is no conventional barrier', claimed the editor of *Physical Culture* in 1900, 'that which was impossible a few years ago is quite correct to-day. It is with the power of every one of you to be healthy and strong.'[164] A series appearing in *The Girl's Empire* magazine two years later, however, marked out the ambivalence that re-emerged regarding girls' health, showing that an emphasis on future motherhood had perhaps been dampened down but it had not vanished, far from it. The feature instructed girls on how to be strong, using Sandow's physical culture techniques. Working 'steadily at these exercises', the girl reader would achieve a sound and healthy body. Yet this strength and health was not tied to personal achievement and individual wellness but to the 'future progress of British Empire': 'When you look at it from an Imperial point of view you see how very important it is. The nation is made up of the individual!'[165] Increasingly, the literature produced for girls aiming to take up sports focused on producing a new generation of robust mothers, with their stockpile of energy 'strengthening them to improve the condition of future generations'.[166]

After Kenealy invented Clara in 1899, for we must conclude that she is an invention, she was referred to by others largely in order to dismiss the dangers of exercise leading to manliness and disqualification from the role of future motherhood. Even Kenealy's eugenicist colleagues seemed more likely to represent the sporting girl as an advantage for the future of the race, provided that she maintained a balance and regard for her future role. In 1912 Amy Barnard described 'bodily fitness in girlhood' as 'the best guarantee of a healthy happy and successful womanhood'.[167] Yet the turn of the twentieth century was also marked by new anxieties triggered by war and a reinvigoration of debates on future mothers of the Empire which cemented collective concerns about the relationship between exercise, health and motherhood, and revived fears about masculine girls. Gordon Stables, encouraged perhaps by the stepping up of Empire rhetoric in the midst of the South African War, urged readers of the *GOP* in 1901 not to 'have too much of that emancipation business'. 'Who wants a woman with biceps, anyhow?... Your women who scull much, or golf or hockey a deal are usually coarse in skin and in features, and far indeed from beautiful.'[168] Whereas his earlier publications emphasised sharing in the sporting activities of their brothers, he now denounced girls' pursuit of 'man-games and tom-boy exercises' that would result in their loss of womanly elegance. The natural order, Stables concluded, was for men to be strong and women beautiful. Stables criticised running and jumping for girls. Golf, meanwhile, led to a 'golf face', 'golf back' and 'the ungainly and hoydenish golf stride', while hockey, was the 'most hoydenish and ungraceful of all man-games and soon gains for her a figure with no more grace in it than of an oyster-wife'.[169] In a turnaround from his earlier advocacy of cycling for girls, he described how biking

> rolls the spine, interferes with the proper function of the hip-bones and gives the bicycle face, with its 'blintering' eyes, look of deep concern, square jaws and flabby mock-turtle cheeks. It is nice to be able to move quickly about in this age of hurry, but biking is after all but a man-game.[170]

As concern with the falling birth rate in conjunction with the high number of infant deaths peaked in the early years of the twentieth century, Mrs Eric Pritchard, wife of the infant welfare reformer and expert on infant nutrition, joined forces with such commentaries, suggesting that the promotion of exercise for girls had gone too fast and too far:

over-development of brain at the expense of body which characterised the early days of women's emancipation has brought about a natural reaction and a swing of the pendulum in the other direction. Just as the cry half a century ago was 'education, education', so to-day the cry is 'physical culture and muscular development for women'; but as was only to be expected, the pendulum has again swung too far in this new direction.[171]

This complex relationship between education and physical exercise will be explored further in the following chapter.

4
Girls, Education and the School as a Site of Health

> E.B. Rheumatic. Heart weak. Gymnastics good for her, but she needs to be carefully watched.

> L.B. Slight and delicate. R. lung not quite sound. Gymnasium very useful but care to be taken.

> F.P. A nervous excitable child subject to headaches. Weak trunk muscles and chest habitually contracted... Must rest between all the exercises longer than the others and not go in when she has a headache or a period.[1]

Reporting in the early 1880s on the health of a group of schoolgirls aged between 11 and 15, Dr Frances Hoggan gave the impression that she was dealing with a poor bunch of specimens. She concluded in her case notes, a few samples of which are noted above, that only one out of five of the girls she inspected had a straight spine, most had poor eyesight, and many of the girls were liable to headaches and anaemia. Defective hearts and lungs were prevalent among the group, as was general weakness and poor physical development. One of the most unfortunate of her subjects was described as 'delicate' and afflicted with dorsal spinal curvature, headaches and varicose veins. Another was described as dirty, her skin in an unhealthy condition, she was small for her age, had an old shoulder stoop and made very 'confined movements'.[2] These examples of ill health and poor physique were far from being under-privileged urban dwellers. Rather, they were the daughters of north London's expanding middle class, day pupils at what was to become one of England's most eminent girls' schools, the North London Collegiate School for Girls (NLCS).

This chapter explores debates on the pursuit of educational goals by young women, an activity which was depicted by many doctors as challenging their already delicate and precarious health. It also examines the ways in which girls' schools were presented – largely in response to these debates – as sites of beneficial health practices. This chapter focuses chiefly on high schools set up for middle-class girls, the subjects of most medical commentary as well as the institutions where the most sustained set of responses to medical critiques took place. It also explores the very real health problems faced by schoolgirls, such as those outlined above. Though these were concentrated amongst poor pupils at elementary schools, they also crossed lines of age and class.

During the late nineteenth century, headmistresses of girls' schools and others supporting the extension of education to young women were compelled to engage with a medical discourse that argued that girls had only limited resources of energy to draw on. Female puberty was defined by many physicians as a challenging epoch in terms of health, which would only be exacerbated by the pursuit of educational attainments. In turn, those promoting the extension of secondary and higher education for girls emphasised the potential for young women to improve their health and vitality to equip them for serious study; schools, by offering a balanced curriculum, were lauded as best placed to support good health practices amongst their charges. This chapter also examines the changing mood at the turn of the twentieth century, when headmistresses, as well as physicians and eugenicists, appeared to re-engage with the notion that overexertion and excessive sporting activity was diminishing girls' femininity and unfitting them for future motherhood. The latter concern, as Pritchard's comments at the close of the previous chapter demonstrate, had never completely dissipated, and many schools presented the preservation of femininity and an emphasis on domestic pursuits as key goals of female education. This rhetoric, however, appears to have been dampened down in the latter part of the nineteenth century as new ideas of girlhood were popularised within and outside of school, and the 'Girton Girl' became emblematic of dynamic, high-achieving girls.[3]

Debates on the relationship between higher education and health were, however, not based on a neat dichotomy between doctors fiercely imposing their ideas about girls' intrinsic weakness and scanty supplies of energy around the time of adolescence, and feminists opposing these arguments. As the proceeding chapters have suggested, there was much ambivalence on this issue amongst physicians and many argued that girls would actually benefit from physical exertion and a good school regime, which had the potential to increase their vitality. Nor were feminists and headmistresses simply driving forward agendas which wholeheartedly advocated higher education for girls; they also expressed reservations about girls' energy resources, gave serious consideration to contemporary medical concerns and worried about the overextension of physical activity amongst their pupils. At the same time, as the first section of this chapter will demonstrate, schools – even those populated by middle-class girls – were forced to tackle very real and serious deficiencies in the physique and fitness of their charges. While what took place in school gymnasia and playing fields was largely about promoting positive health and physical culture, headmistresses, gymnastics teachers and a growing corps of women medical superintendents and inspectors devoted a great deal of time to tackling what could best be described as remedial health issues.

Considerable attention has been devoted in the secondary literature to the evolution of exercise cultures and sporting activities in girls' schools and women's colleges. While the previous chapter demonstrated that the extension of exercise cultures and practices for girls took place in a wide variety of settings, women's university colleges and secondary schools have been credited with releasing young women, by means of sport and exercise, from their curtailed biological and social roles.[4] Paul Atkinson has suggested that while 'Women's emancipation was not won on the playing fields of Roedean...the games fields and gymnasia of the girls' schools were the scene of an important skirmish in the struggles of Victorian feminism.'[5] This chapter will develop a somewhat broader perspective, examining the myriad approaches adopted by schools to improve girls' health, referring not only to gymnastics, sports and games, but also to the ways in which health was tackled as part of a broadening curriculum, and links made between school and home in health-promoting activities. Schools were presented as having the potential to counter the impact of poor environments, home conditions and training, while education in health matters became a key aspect of the curriculum during the late nineteenth and early twentieth centuries. The expansion of girls' schools also created multiple opportunities for women doctors in the role of school medical inspectors and several, like Frances Hoggan at the NLCS and Catherine Chisholm at Manchester Girls' High School, were influential within and outside their institutions, publishing on girls' health and exercise and lecturing regularly on this topic.

Schools as sites of remedial medicine

Many girls had a very poor starting point for any programme of activity intended to promote good health and physical training. This was particularly notable in elementary schools, which dealt with more fundamental challenges in terms of the health of their pupils. When, in 1887, Martina Bergman Österberg, then Superintendent of Physical Education to the London School Board, gave evidence to the Cross Committee investigating the working of elementary education in England and Wales, she expressed grave concern about the physical training of girls and the absence of facilities in schools, which meant that much drill ended up being 'desk drill', with an emphasis on discipline rather than physical development. Headmistresses were also liable, she argued, to introduce systems of their own, which were 'deficient and even injurious through want of anatomical and physiological knowledge'.[6] Bergman Österberg also agreed with the Committee's proposition that 'there are a few schools in London, are there not, in which you would describe a large number of the children as being in a state of semi-starvation?' In these cases, she recommended moderated gymnastic lessons. In the very poorest schools, where muscular exertion would take too much out of the children, she suggested that exercise should be abandoned altogether.[7]

Bergman Österberg was eager to introduce the Swedish or Ling system of 'free exercises' to London's schools, and by 1887 the London Board had 1,200 teachers (one-third of all schoolmistresses employed) trained in the system, which had been introduced to 250 elementary schools. In 1898 it was claimed that 'there is scarcely a child attending our [London School Board] schools who is not receiving systematic physical training', using the Swedish system that did not require any apparatus[8] (see Figure 4.1). By this time, however, Bergman Österberg had resigned from the London School Board. Frustrated by the terrible physical condition of many London pupils, she argued that this made the introduction of effective physical training extremely difficult, and acknowledged that 'her ideal of healthy womanhood could not be realized among the poor'.[9] Yet the Swedish system continued to expand, and after 1909 the Board syllabus included step marches, dancing, skipping and gymnastic games. During the early years of the twentieth century, however, female school inspectors continued to criticise the lack of space, insufficient emphasis on movement, free play, exercise and fresh air in schools, compared with the continued emphasis on discipline and military drill.[10]

That ideas on improving the health and physique of children had broader currency is exemplified by the activities of the Froebel Society. Set up in 1874 to encourage kindergarten training, the Society was also influential in propagating broader ideas of remedial treatment in childhood, to restore children crushed by capitalism and the city to nature.

FIG. 7.—BALANCE MOVEMENT.

Double heel raising and knee bending, executed
with hands at "hips firm."

Figure 4.1   Leg and balance movement (note the poor environment in which exercise took place, and the oversized dress worn by the girl in the photograph). Mrs Ely-Dallas, Organising Teacher of Physical Exercises under the School Board for London, 'Physical Education in the Girls' and Infants' Departments in the Schools of the London School Board', in *Special Reports on Educational Subjects*, vol. 2 (London: Her Majesty's Stationary Office, 1898), Figure 7 (author's collection).

Margaret McMillan's pioneering initiative in setting up a Camp School in Deptford around 1910, meanwhile, was 'intensely practical' in its offering of minor treatments, such as the removal of adenoids, the promotion of healing through remedial gymnastics and provision of meals in its garden setting to intensely deprived local children.[11] In 1890 the London School Board appointed the first school medical officer in England, and after 1908 the health of the schoolchild was placed within the more rigorous framework of the School Medical Department, which eventually provided a number of basic medical treatment schemes.[12] Yet one school medical survey of 1906 alluded to the fact that young children who survived the slums could be sturdier than 'more carefully nurtured children', who were often 'coddled and pampered, and lymphatic conditions from want of exercise, enlarged tonsils, adenoids and anaemia are essentially the troubles of these better-class schools'. Overall, however, girls tended to have lower standards of nutrition, and to be more liable to anaemia and poor eyesight as they spent less time outside than boys, undertook a larger proportion of domestic work, took less exercise and their eyes were strained by close needlework.[13]

At around the same time that Bergman Österberg was reporting to the Cross Committee and Frances Hoggan was commenting upon the poor physical state of her charges at the NLCS, London gymnastics teacher Miss Chreiman was highlighting – striking what appeared to be a contradictory note – the importance of physical culture for 'the quality of our girl-life ... for certain is it that upon the moral and physical qualities of our women depend mainly the security, order and happiness of our households, and the physical, intellectual, and moral improvement or deterioration of our race'.[14] She went on to assert that education went hand in hand with health: girls, as well as teachers and parents, needed to be encouraged 'to get into the modern scheme of female education'; girls were enjoined to improve their vitality, 'and understand that the "conjugating of a Greek verb is as much a physical process as is the walking on a tight rope"'.[15] She urged physical educators to work closely with medical advisors and mothers to improve the health and vitality of the girls in their charge. Miss Chreiman's lectures were stirring and inspirational, as she described her longing for a kind of *physical religion*, a 'power for energizing and elevating soul and body'.[16] On a more pragmatic level, as part of her mission to promote physical education, she carefully laid out programmes of work in the gymnasium, adapted for particular parts of the body and particular conditions of individual girls.[17]

The contrast between Chreiman's account of what girls' health *should be* and Hoggan's assessments of what girls' health actually *was* appears stark. However, the two women were bound together in time, space and purpose. Frances Hoggan was appointed Medical Superintendent to the NLCS in 1882, with the remit of acquainting herself with the physical condition of each pupil and monitoring their progress. She was to report cases of physical weakness to the headmistress, Miss Frances Buss.[18] At the same time, Miss Chreiman was invited into the NLCS to train its pupils in gymnastics and to liaise with Hoggan on individual programmes of gymnastic work designed to match the pupil's physique and abilities. Chreiman's work was greeted enthusiastically by Frances Buss. Indeed, working from her base at the Institute of Physical Culture in Baker Street, Miss Chreiman and her lady assistants enjoyed widespread acclaim for combining the best of gymnastic systems into 'an arrangement of varied and graceful exercises eminently calculated to attain the end in view'. Her fan base included Edwin Chadwick, an advocate of Ling's gymnastics; he recommended Chreiman's system as a way of relieving the 'exhaustion' and 'bodily languor' of book and desk work, and urged its adoption in every school in the Empire.[19] The periodical *Health*, meanwhile, praised Miss Chreiman for developing an effective and pleasurable system of gymnastic exercises for girls, intended to develop bodily strength, correct 'defects of carriage' and encourage healthy growth.[20]

Miss Chreiman's spirit of optimism and Dr Hoggan's concerns about the health status of the girls in her charge closely dovetailed. Both believed that physical training was a necessary precursor of study, and regretted its neglect for girls around the time of puberty. Hoggan outlined how differences in strength and bone structure between boys and girls became accentuated from the age of seven, with girls showing marked tendencies to spinal weakness, deformity and stooping, expansion of the pelvis with less developed chests and muscular systems. Physical training was intended to resolve these problems.[21] Hoggan, interested in preventative approaches to health more broadly, commented how

> From the time of their regular entrance into the schoolroom, they are expected to lay aside all vigorous play, and to be a hoyden or a tomboy is often thought to be the very acme of impropriety in a young schoolgirl. Intellectual training in the better class of schools, dull learning by rote in the inferior ones, takes henceforth the first place in girl's education, and seldom indeed do we find the physical needs of a growing and delicate organization come in for anything approaching adequate attention either in school or home education ... with girls especially, the training of the physical powers should take precedence over the training of the intellectual powers, the latter being incapable of unfolding harmoniously in a stunted or deformed body.[22]

Both believed that girls had enormous potential for improvement, yet recognised that great patience and care was needed when introducing them to physical training and that the starting point in terms of health, physique and posture was often poor.[23] Chreiman cautioned that 'The exercise of beginners should be conducted with the greatest method and care; the general capacity of the pupil should be ascertained, and any local weakness well considered.'[24] Though their intake was largely middle class, the NLCS accepted girls from a variety of social backgrounds (and it was not unusual for girls to be taken out of the school if family incomes diminished) and, while the nutritional standards and health status of its pupils were likely to be very much in advance of those attending state elementary schools, the poor health of many of them was striking. Consequently, emphasis was placed on individually designed programmes of work and correctional exercises. Miss Chreiman described how, in general, much of her work with girls was remedial, addressing eye and ear defects, muscular weakness, poor deportment, coordination, lack of sleep, tendency to disease, and skin and heart conditions. She urged that

> exercise should not be too violent in character, and that while all parts of the muscular system are sufficiently exercised, no part or parts assume under preponderance of work or power. The truest and best results are brought about by moderate persevering efforts spread over a long period of time.[25]

This approach reflected the view of head teachers, teachers of gymnastics and doctors interested in the health of schoolchildren more broadly and concerned about girls' poor physical status. A. Alexander, Director of Liverpool Gymnasium, explained that much of his work was dedicated to remedying defects and strengthening weakly frames; he was particularly preoccupied – as were many of his contemporaries – with the problem of spinal curvature, the result of too much desk work or want of exercise when young: 'If girls went through a gentle course of apparatus drill while at school, the danger of spinal curvature would be considerably lessened, and the whole physique strengthened and shaped to its natural formation.'[26] Exercises utilising French bar bells were recommended for delicate girls, the chest-machine for weak-chested girls, and travelling rings were described as being particularly useful 'in cases of lateral curvature, in promoting an equal development of the shoulder blades and spinal muscles'.[27] Dr Clement Dukes, author of a number of books and articles on the relationship between health and schooling, outlined the ways in which suitable exercise, as

well as attention to seating and lighting, helped develop straight backs
as well as remedying other defects such as knock-knees and outgrow-
ing shoulder blades.[28] He condemned stays and high-heeled boots, and
warned against the practice 'of requiring lessons to be *written out*'; 'the
pupils get so fatigued that they lounge over the table in consequence'.[29]
At the end of the century, Jane Frances Dove, headmistress of High
Wycombe School, was still emphasising the importance of gymnastics,
which allowed for close observation of individual development:

> since every girl must appear twice a week in the gymnasium as long
> as she remains in the school, dressed in an easy fitting costume,
> consisting of knickerbockers and tunic, the gymnastics mistress has
> every opportunity of noticing the physical development, and I have
> found that she very quickly detects even the slightest curvatures or
> other physical defects, and, with the parents' consent, can give cura-
> tive treatment, which is very speedily efficacious in curing weak or
> crooked backs, stoops, displaced shoulder blades, sprains and other
> ailments.[30]

In 1879 Frances Buss pioneered a scheme of individual medical inspections for the NLCS girls, related to directed gymnastic work and exercise
and, outside of school, regulated homework and physical activity,
negotiated through letters to parents. The medical inspections were conceived as a direct response to concerns about the impact of education on
growing girls, but were additionally intended to discover health problems and weaknesses before they became entrenched. They involved the
collection of detailed data on height, weight, chest measurement, lung
capacity, lifting power, head measurements, sight, hearing and 'special
peculiarities'.[31] The duties of the medical superintendent were carefully
outlined, and in particular they were to be responsible for establishing
a close acquaintance with the condition of each pupil, watching for
signs of overpressure and devising programmes of individual gymnastic
exercises. Many of the NLCS pupils were excluded from certain activities due to poor health, and work in the gymnasia was interspersed
with rest periods. Frances Hoggan described only six out of 335 girls
in her charge as strong and vigorous; the remainder had spinal curvature, ruptures, acne, knock-knee, lung congestion and heart disease. One
14-year-old girl, seen at Hoggan's first medical inspection, 'has had rick-
ets and now has attacks of momentary stupour, especially when at all
pressed to work...Will gain much from the gymnasium, but must be
treated...as a child of eleven or twelve.'[32] H.R., examined by Hoggan

in June 1883, was reported to be suffering from dorsal spinal curvature, frequent headaches and general delicacy, attributed to a bout of peritonitis. She was allowed to do exercises with her left arm, but not her right, and she was not to undertake the giant stride. By October she had improved significantly and was permitted to do all the exercises with both arms.[33] O.P. was reported to be gaining in strength but was still advised to rest after tiring exercises. 'She ought at home to swing from a pole by both hands for 3 minutes night and morning in loose clothing, in order to strengthen her back. She has headaches, and the duration of home work ought to be inquired into.'[34] F.H. was diagnosed with spinal problems and had a prominent shoulder blade, but she had a well-developed trunk, which was in part attributed to the fact that she did not wear stays. She was given suspension exercises to do at home and a moderate programme of work in the school gymnasium. Two years on in 1884, at the age of 16, she was reported to be 'worse, posture habitually bad', pale and delicate and working too much at night.[35] What became a much reproduced image of the NLCS gymnasium in action shows the girls engaged in a range of swinging and stretching activities, as well as the 'giant stride', designed to deal with their curved backs, sagging shoulders, lopsided gaits and weak grasps. One girl is shown resting between exercises. Many girls were forbidden to participate in the giant stride and, despite the school's best efforts, remained in a poor physical state, and were treated for the purposes of physical training as if they were two or three years younger (see Figure 4.2). Health and exercise were serious matters for headmistresses, teachers and school physicians, and involved dealing with girls with grave conditions. Girls were frequently sent home sick, and in 1882 the NLCS magazine reported the deaths of two 15-year-old girls who had recently left the school.[36]

Medical theory, balance and education

Though headmistresses, physical educators and school medical officers became preoccupied with the 'realities' of pupils' poor health and physique – as well as the promotion of girls' schools as an ideal site to improve the health of their charges – medical commentators focused more persistently on the impact of study, 'brain work' itself, on girls' wellbeing, nerves and reproductive organs. While the Education Act of 1870 had signified a major extension of the national public education system, and the 1902 Act saw public secondary education become increasingly accessible for girls, the debate on female education tended to focus on the schooling of middle-class girls, those most likely to

THE GYMNASIUM OF THE NORTH LONDON COLLEGIATE SCHOOL FOR GIRLS.

Figure 4.2   The North London Collegiate School Gymnasium.
Source: From E.A.L.K., 'The North London Collegiate School for Girls', *Girl's Own Paper*, III:
122 (29 April 1882), facing p. 494 (The British Library).

follow up secondary education with a university degree or work in
the professions. As seen in Chapter 1, debates on the impact of brain
work for girls built considerable momentum following the publication
of Dr Henry Maudsley's article 'Sex in Mind and in Education' in 1874.
Maudsley's piece was primarily concerned with the increased provision
of secondary and higher education for young women. He predicted a
gloomy future for high school and college girls, unable to bear 'excessive mental drain as well as the natural physical drain' during their
studies, which could lead to an 'imperfectly developed reproductive system', mental, moral and intellectual deficiency; 'the individual fails to
reach the ideal of a complete and perfect womanhood'.[37] Maudsley did
not deny that girls could become good scholars and pass examinations,
but they would do so at the expense of their physical and reproductive development, would lose their natural grace and femininity,
and become imperfectly developed creatures, liable to produce degenerate offspring or render themselves infertile. 'It is quite evident', he
declared, 'that many of those who are foremost in their zeal for raising the education and social status of woman, have not given proper

consideration to the nature of her organization, and to the demands which its special function make upon its strength'.[38]

Prompted by her fellow supporter of universal suffrage and educational reformer Emily Davis and headmistress Frances Buss, Dr Elizabeth Garret Anderson produced a robust challenge to Maudsley's assertions.[39] Then a governor of the NLCS as well as a member of the London School Board, Garrett Anderson asserted that it was

> a great exaggeration to imply that women of average health are periodically incapacitated from serious work by the facts of their organization... With regard to mental work it is within the experience of many women that that which Dr. Maudsley speaks of as an occasion of weakness, if not of temporary prostration, is either not felt to be such or is even recognised as an aid, the nervous and mental power being in many cases greater at those times than any other...

> The case is, we admit, very different during early womanhood, when rapid growth and the development of new functions have taxed the nutritive powers more than they are destined to be taxed in mature life... [but] the assertion that, as a rule, girls are unable to go on with an ordinary amount of quiet exercise or mental work during these periods, seems to us to be entirely contradicted by experience.[40]

Garrett Anderson pointed out that life at a good school, with provision for fresh air and exercise, was for many girls better than staying at home, where their health was broken down by boredom and dullness, and that education for women was not just a matter of equipping them for work, but also a time to cultivate reason, conscience and imagination. She concluded that physical education was especially necessary for girls, and also pointed to the careful management that was required to steer girls though schooling as well as a healthy puberty.[41] By the time these essays were published, Maudsley was already swimming against the tide. While the debate on educational over-pressure and its impact on girls' physiological development would reverberate in the medical and public press for several decades, practice quickly overtook discourse. The main expansion in endowed and propriety schools for girls occurred after 1870, and some 200 such schools were set up in the second half of the nineteenth century, although in 1900 most middle-class girls schooled outside of their homes were taught in the more traditional, privately owned schools. Following the 1902 Education Act, the number of girls educated in grammar schools rose significantly, and many less affluent middle-class girls received an academically orientated education.[42]

As opportunities for girls' education expanded from the mid-nineteenth century onwards, headmistresses placed a great deal of emphasis on achieving a balance between mental and physical training. A number of them continued to advocate 'accomplishments' to counter girls' mental exertion because they did not 'wear the mind' and indeed held 'an analogous position to the cricket and foot-ball of a schoolboy's life', but increasingly emphasis was placed on the benefits of exercise and sport.[43] Contributors to the 1868 Taunton Report on the education of girls, including Emily Davies, Dorothea Beale and Frances Buss, criticised girls' schools for their feeble attempts to introduce physical exercise, which involved a good deal of 'listless idling' and 'the dreary two and two walk'. Girls lacked 'the vigour and joyousness which belong to the free and healthy play of boys'. One commentator argued that girls needed to play games with 'zest' and '*abandon* and self-forgetfulness'. 'If their play were more bracing and recreative, their mental improvement would be sounder and more rapid.'[44] Mr Fearon cautioned that mental pressure could injure girls, but that this was aggravated in many instances by a want of early training, neglect of physical education, and an absence of diverting exercise in many schools.[45] Headmistress and champion of women's rights, Elizabeth Wolstenholme, claimed that girls worked through examinations with little difficulty 'without apparent fatigue or ill effects whatsoever' and denied reports of frequent fainting and hysterics.[46] Headmistress Dorothea Beale described how study was limited to four hours a day at Cheltenham Ladies College, interspersed with music, drawing and exercise. Girls were expected to do a limited amount of homework, and she had no evidence that school work was injurious in any way.[47] The report also commented on the lack of provision of playgrounds and for open air exercise and games in many girls' schools.[48]

The feminist movement and campaign for suffrage, as well as women's entry into medicine, formed a crucial backdrop to changing notions of health and its relationship to school work. In 1866 Emily Davies explained that 'it is a rare thing to meet with a lady, of any age, who does not suffer from headaches, languor, hysteria, or some ailment showing a want of stamina... Dulness [*sic*] is not healthy, and the lives of ladies, are, it must be admitted exceedingly dull.'[49] Davies saw the extension of higher education for women and the encouragement of good health as feasible objectives, providing a route out of this impasse. Dr Elizabeth Blackwell, the first woman to be listed on the Medical Register in England and, together with Hoggan, co-founder of the National Health Society, with its aim of promoting sanitary education, roundly condemned the state of physical education in girls' schools in her *Lectures on the Laws of Life*, published in 1871.

> If we were to sit down and carefully plan a system of education, which should injure the body, produce a premature and imperfect development of its powers, weaken the mind, and prepare the individual for future uselessness, we could hardly construct a system more admirably calculated to produce these results.[50]

Blackwell called for a more enlightened approach to physical and mental training, yet tempered this with concerns about the dangers of too rigorous an intellectual engagement, which threatened to wreck girls' reproductive health and marriage prospects, lead to a neglect of their domestic training, and imperil the health of individuals and the future of the race. She advocated a focus on healthy practices and moral education coupled with accomplishments, including piano and embroidery, and restraint in imparting technical knowledge, until girls reached the age of 16: 'It is of much greater importance to her, that she should possess a strong, straight back, good digestion, a cheerful temperament, a body that can move with vigour and grace, organs that perform their functions healthily.'[51] Blackwell's caution was shared by other female doctors writing on this theme and even headmistresses such as Buss, who was keen to prepare her pupils for any position in life, would not deny the importance of traditional feminine roles and domestic knowledge.[52]

Despite the mediating tactics employed by early feminist doctors, male practitioners, notably those trained in psychiatry like Henry Maudsley, maintained an oppositional stance to female education, a stance no doubt prompted in part by their own anxieties about the incursion of women into their field of practice as well as their contact with patients suffering from what they considered to be the effects of over-education. Others, however, put forward more nuanced views. In 1880 Edinburgh psychiatrist Dr Thomas Clouston declared that he was certain that every brain had only a 'certain potentiality' for education in one direction and for power more generally, and regarded adolescence, rather than puberty, as the most dangerous time, especially for young women, given that by this time the 'functions of reproduction' were more fully developed.[53] In 1882 Clouston published *Female Education from a Medical Point of View*, later reprinted in *Popular Science Monthly*, which described the long hours of study and fearsome competition in more advanced schools for girls, which left pupils suffering from anaemia, stunted growth, nervousness, headaches, neuralgia, hysteria and insanity, and stimulated morbid cravings for alcohol or drugs, as well as affecting women's highest function of motherhood. Yet the bulk of his article dealt more broadly with careful monitoring of girls and ideas on how to temper and adapt school regimes: 'If a girl has grown a couple of inches a year, then depend upon it she should not study hard... You want not only growth, but activity, grace of movement, alertness, strength. You won't have these if the girl goes on studying hard while she is growing fast.'[54]

Dr John Thorburn, Manchester Professor of Obstetrics, warned in 1884 that women moving into higher education were entering a dangerous occupation, which could lead to complete unhappiness, particularly if they did not rest during their periods.[55] Thorburn urged 'calmness and impartiality' – as 'a professor of a college and a university which are perhaps doing as much as any in England to meet the demand for female degrees... I would not for a moment be supposed to join in the ignorant clamour against such education, which is based on a desire for monopoly by the male sex.'[56] But he cautioned 'up to this time, no means have been found which will reconcile this with the physiological necessity for intermittent work by the one sex'. It was the duty of 'every honest physician to make no secret of the mischief which must inevitably accrue, not only to many of our young women, but to our whole population, if the distinction of sex be disregarded'.[57] He was also quick to report on the death of one of the first female students in Manchester from tuberculosis, arguing that her death resulted from 'over-education'.[58]

While Clouston described the later years of adolescence as a time of particular risk, other doctors claimed that it was the establishment of periodicity itself that was especially disabling, a phase that should be marked by complete abstinence from schooling or any otherwise intellectually demanding activity; instead, girls should rest on sofas or retreat to the countryside. Others merely advised caution and urged schools to promote periods of rest or breaks from teaching when girls had their periods. In the mid-nineteenth century Edward Tilt had suggested to mothers practical measures to delay the onset of menstruation, while also broadly supporting a rounded education for young women.[59] In a letter to the *British Medical Journal* (*BMJ*) in 1887, however, Tilt proposed that even if menstruation was normal, girls should be prohibited from cramming for examinations or 'working out complicated problems of arithmetic'. John Thorburn claimed that he had personally seen 'a very large number of cases where the growing schoolgirl had been temporarily or permanently injured by scholastic work during the menstrual period'. Typically these girls were anaemic, hysterical and suffering profuse or absent menstruation, intense periodic discomfort or more general disorders of the nervous or digestive system.[60] He blamed this not on the conditions at schools, which tended to be better than at home with regard to diet and hygiene, but schools failed when it came to considering the 'menstrual week'. Girls, meanwhile, would soldier on and break down at the end of term, resulting in grave disease, uterine displacement, pelvic disorders and mental breakdown, or even death.[61] Thorburn accorded a particular role for mothers in the management of puberty; if they learned to manage their daughters' mental efforts during menstruation, this, he argued, would impact on schools, who would adjust 'their arrangements to permit of an average rate of general progress which will allow for an average loss of time by each individual girl'.[62] Despite the tone of his warnings, Thorburn still stressed that girls' education was a matter of adaptation rather than prohibition and, by the early twentieth century, medical commentators such as Howard Kelly were remarking that 'the amount of menstrual disturbance among school girls has been much over-estimated'.[63]

As Emily Pfeiffer was to discover in compiling material for her book, *Women and Work*, published in 1888, it was not difficult to collect testimony favourable to women's education, even from medical men. She organised her study into separate chapters on physiological-medical evidence opposed to higher education for women and medical and statistical evidence which was broadly favourable and generally asserted the need for attention to be paid to the girls' regimen and environment at school.[64] Sir William Gull suggested that girls needed their food, air, exercise and sleep to be attended to and that 'gymnastics' of the mind and the body should be limited 'within the individual's natural power'. But he also suggested, despite his views on anorexia, the psychological deviance that it entailed and its frequent association with over-achievement, that 'In the light of medical experience, the advantage to health of good and even high intellectual training for girls and young women cannot be doubted.'[65] Dr Hermann Weber believed that girls who devoted themselves to 'judicious intellectual work' were unlikely to suffer more than those who follow 'the ordinary pursuits of fashionable society', describing how

> Now and then I have met in my professional work with girls and young women who had injured themselves while working together at college and at home. In almost every one of these cases, however,

the illness was due to mistakes committed in studying than to the intellectual efforts *per se*. Such mistakes, for instance, were want of sufficient air and exercise; injudicious arrangements as to meals and sleep; imperfect ventilation of bed-rooms and sitting rooms; over-anxiety about the results of examinations.[66]

A balanced take on the subject was laid out in 1895 in the pages of the *BMJ* by gynaecologist and 'nerve doctor' Dr William Smout Playfair. He urged, in considering the education and training of girls 'of the easy classes', the need to 'bear in mind ... the sensitive nervous organisation of the human female ... but much more especially diseases of the reproductive organs'.[67] At puberty 'the conduct of the health of the growing girl may influence for good or evil the whole future of the woman'. He warned against girls being raised 'like hothouse plants', which would lead to loss of stamina, producing 'the wretched broken-down invalid so often met with in the present day, especially in the upper ranks of life'.[68] As higher education for women had taken such enormous strides, it was important, Playfair argued, to discuss such issues and to avoid 'the unfortunate tendency on the part of many mistresses of high schools to listen to the warnings of medical men with incredulity, to accuse them of narrow-mindedness and opposition'. The problem was the following, Playfair argued. Girls' schools failed to devote sufficient attention to the establishment of menstruation and to make provision for physical activity, and girls were given unreasonable amounts of work.

> As evidence of this ... I have been consulted in the case of two young ladies, aged respectively 11 and 16. One was chlorotic, and her menstruation had ceased for a year. On taking her timetable at a well-known high school, I found she had 7¾ hours of work, an amount not in itself, perhaps, excessive in a healthy girl. From 2.30 to 4 there were no lessons, and, if the weather permitted, she might if she liked take a walk, but it was not insisted on, and being naturally languid and listless, as all such cases are, she rarely did so. There was no other opportunity for exercise at all.[69]

The other suffered from menorrhagia, anaemia and debility. Her timetable took up seven to eight hours daily and there was no provision for exercise. In neither case, Playfair asserted, had the school attempted to inquire into the girls' menstrual function or to adapt the course of study to their needs. Girls' schools needed to provide more exercise for girls – including rowing, golf or cycling – to make games compulsory, and to abolish corsets. This 'would do more towards lessening the number of neurotic women that the medical profession has to deal with than the medical profession can possibly do by any exercise of its own art'. It should also be the 'bounden duty' of headmistresses, parents and doctors to insist on the cessation of severe study if girls showed signs of illness, menstrual disorders, chlorosis, wasting or loss of appetite.[70]

As Kate Flint has asserted, what was new and important about the discussion generated from the 1870s onwards about the relationship between health and female education was 'the degree to which it spread from specialist texts to a range of publications with the potential to reach a wider readership'.[71] The fact that Maudsley and Garrett Anderson opted to place their views directly into the public domain by publishing in the *Fortnightly Review* was highly significant. The debate demanded and won energetic responses from the headmistresses of girls' schools and in the press and periodical literature, prescriptive literature, health journals and girls' magazines, triggering discussion of how girls were to be equipped with the vitality necessary to tackle more structured academic curricula and examinations, drawing on British and American examples of girls' experiences of school and college life, as well as the new exercise regimes and sporting possibilities opening up to young women.

Not all accounts were positive, far from it. In May 1894 *Good Health* published a critical account of the impact of education on American girls, describing how 'in hundreds of homes bright and interesting girls have been fatally injured by indiscriminate and unwise work'.[72] A month later the journal underlined this assessment, attacking the expansion of female education which attempted to make women more like men: 'Nature has put so many obstacles in their way that they can never succeed to any great extent, but the effort does harm to health and character.'[73] In 1896 the viewpoint of psychiatrist Sir James Crichton-Browne on the education of women was also reported in *Good Health*.[74] The article suggested that his opposition was driven by a concern that it would destroy the physical beauty of woman, alluding to Crichton-Browne's claims that the female brain was fundamentally smaller than that of the male and that mental illness was common in overworked schoolgirls. 'The aberrations may be slight ... but now and again they advance into that wild coma which corresponds with apathetic dementia.'[75] Over-pressure could also lead to anaemia, general delicacy, bad complexion, short-sightedness, and loss of hair and teeth, 'all fatal to womanly beauty'. Summing up 'Troubles ahead for women', the article concluded

Over-pressure from ten to seventeen years of age may have amongst the remote consequences not only the reproduction of the functional nervous disorders which so often manifest themselves at that period, but a crop of gross nervous degeneration which has up to this time been rarely seen in women.[76]

Crichton-Browne's lengthy chapter on education and the nervous system in Malcolm Morris' *Book of Health* also warned in no uncertain terms about the dangers of continuous study for girls, which, he argued, was likely to trigger nervous derangement and mortal disease. Schooling could only have negative effects if not particularly adapted to the 'female organisation', leading to brain exhaustion, bodily infirmities and phthisis, though he acknowledged that fresh air, muscular work and outdoor occupations could be beneficial. He regretted the passing of the age when girls 'of the more affluent classes' were initiated into 'useful household duties...which imposed no strain on the intellect, and but little ruffling of emotion'.[77] Crichton-Browne acknowledged that schooling under incorrect conditions could impact badly on boys too, who could suffer from the adverse effects of brain exhaustion, but never to the same extent as girls.

Crichton-Browne's medical papers, meanwhile, presented evidence on the substantial variations in the size of male and female brains, as well as their structural differences, which made women more emotional and volatile, and he linked the high incidence of phthisis in young women to over-pressure at school. He also alluded to the doubts expressed by high-school mistresses themselves by 1892 about over-pressure and too close an application to study: 'it is terrible to contemplate the ruin that might be wrought amongst them were they to be ruthlessly urged on and subjected to competitive excitement', but at the same time Crichton-Browne brought no 'wholesale indictment' against high schools for girls, claiming that they had done 'good service to sound education' and 'supplanted second-rate boarding schools and venture schools, which were hot-beds of namby-pambyism, and have opened up to girls interests and helpful attainments which were formerly denied them, thus saving some of them from a vapid and weary existence'.[78]

The topic of 'overexertion' or 'overwork' continued to command a great deal of attention. 'Girls', declared Dr Clement Dukes, physician to Rugby School and author of a number of books targeted at educationists as well as broader publics, were 'naturally more subject than boys to nervous excitement'. This, however, could be 'effectively restrained by sound physical development'. He blamed 'faulty training' for girls being ' "nothing but nerves" ... "nothing but emotions" '; these evils would disappear under 'a more reasonable system'.[79] 'They are simply suffering from their faulty education, their narrow mode of life, and their preposterous nineteenth century dress. We want more muscles and less nerves in our girls, if we are to have them healthy and vigorous.'[80] Dukes condemned girls' schools for their want of attention to physical education, and he was at pains to emphasise the damage resulting from over-pressure, strenuous examination systems, too long a school day, inattention to the environment of the school and poor diet, which wreaked havoc on the health and development of both boys and girls. Overwork, Dukes claimed, resulted in restlessness, inertness, loss of originality and interest, a sallow appearance, sleeplessness, flabby muscles, headaches, stammering, a tendency to shirk society, wasting, constipation, poor blood circulation, hysteria, brain fever and mania. 'Looking to the number of hours some children, especially girls, are compelled to work, their brains must of necessity be pretty well addled', and for girls overwork could in addition lead to disorders of the generative functions.[81] Dukes cautioned that great care was required to ensure that girls were not subject to competitive examinations, which could lead, given that they were naturally more subject to nervous excitement, to breakdown.

However, Dukes also emphasised the potential role of the school in shaping and improving girls' health prospects rather than damaging them. The school needed to allow for sufficient rest for younger pupils, particularly between the ages of 11 and 14, when, according to Dukes, the girl 'suddenly develops into a woman'. Overall, girls required as much exercise as boys; this could include skipping, tennis, swimming, riding, rowing, dancing, drill and gymnastics, running and skating. Daily exercise would eliminate conditions such as curvature of the spine and consequent ill health, and schools were urged to provide chairs with proper back support, rather than backboards, to produce straight backs.[82] He pointed to the importance of hygiene and morning cold baths for girls, and the dangers of defective clothing, including high-heels and stays. Elsewhere, he elaborated on the diet necessary for growing schoolgirls, cautioned schools not to permit them to stint themselves and to provide sufficient food for rapid growth.[83] The school, in Dukes' view, was a potential site of health that could improve the health of young women as well as making them smarter, rather than a place which endangered girls' health and future capacity for motherhood.

Headmistresses also engaged actively with the question of overwork and, in addition emphasised the issue of under-achievement, which they perceived to be endemic and equally bad for girls' health. Dorothea Beale had urged a robust and bracing approach to girls' education in a paper delivered to the Social Science Congress in October 1865.[84] She argued – her words in many ways prefacing those of Elizabeth Garrett Anderson – that work, study and occupation were vital (poor girls learned this automatically) and that

> for one girl in the higher middle classes that suffers from over work, there are, I believe, hundreds whose health suffers from the fever- ish love of excitement, from the irritability produced by idleness, frivolity, and discontent . . . I am persuaded, and my opinion has been confirmed by medical doctors, that the want of wholesome occupa- tion lies at the root of much of the languid debility of which we hear so much after girls have left school.[85]

The theme of overwork was taken up by Sophie Bryant, Frances Buss's successor as headmistress of the NLCS, who urged teachers to mount a more robust response to bring 'counter-charges against our friends, the doctors', and to allay the 'panic' around the theme of overwork.[86] Bryant pointed out not just the challenges of over-zealous, bustling girls, but also the 'grievously large number who are over-balanced on the side of indolence'. Such girls liked nothing better than reading novels 'which keeps up a pleasant flow of ideas and images with a pleasurable sur- rounding of emotional excitement', but were far less keen to engage with Latin, mathematics or music. Other girls erred on the side of phys- ical activity and yet another group was highly emotional and liable to nervous exhaustion. In the middle was the girl who was well balanced in body and mind, who enjoyed all kinds of physical and mental activity, who Bryant regarded as a golden mean to be aspired to.[87] A mathemati- cian herself by training, Bryant advocated a balanced curriculum and monitoring the girls' performance, but also 'We know that the one thing necessary for the girl's moral and, in the end, physical health is that she should be induced to work somehow.' Girls who were thoroughly idle and who scarcely, if ever, made any mental effort, she argued, were much more likely to end up under the treatment of a doctor.[88]

Girls' schools and practices of health

Paul Atkinson has argued that sport and exercise in schools was har- nessed largely as a defence tactic against accusations that girls' schools were wrecking their pupils' health, and implicitly their future reproductive capacity.[89] This is likely to have been one of the drivers for the approaches to health and physical training adopted by girls' schools. However, as well as being conceived as a response to negative publicity and ongoing debates about the risks of female adolescence in general and of schooling in particular, school practices were shaped by very real anxieties about the poor health of their pupils. Schools began to present themselves as sites of good health practices and safe havens from damaging home environments and poor training. However, the move into a sphere that was traditionally the responsibility of parents, particularly mothers, in the home – and, many would argue, a neglected responsibility – had to be managed with diplomacy. Frances Buss at the NLCS certainly moved beyond a defensive strategy when developing a programme of health education which urged girls to be good all-rounders in terms of their physical and mental health.[90] Buss saw the NLCS as a progressive environment, where the best health care could be employed, and new techniques and regimens of health introduced, with the school working in cooperation with parents. Headmistress of the NLCS since its inception in 1850, Buss steadily introduced a regime of directed gymnastic work and games, and paid close attention to the nutrition, hygiene and health of her pupils and to the school environment. In 1868 she reported to the Taunton Commission that 'lessons on the structure of the human body, with applications to health' had been introduced, and emphasis placed on bodily exercise, which was vital for mental exertion.[91] By 1879 Buss had introduced a prize for the best essay on the laws of health and had adapted the school hours to allow more time for callisthenics.[92] Swimming at the St Pancras baths, hockey, tennis, dancing and games were introduced, all closely regulated, but it was – as we have seen above – graduated gymnastic exercises that were seen as the linchpin of girls' physical development, and after the construction of the gymnasium every girl was to have two gymnastics lessons a week and drill on the other three days.[93] Special attention was paid to diet, with the school offering plain dinners for 10d to its day pupils. The NLCS had an ambitious academic ethos, and encouraged its pupils to fit themselves to earn their own living, to strive for university education and careers in the professions, and to take competitive examinations, including the Oxbridge entrance exams. Exercise was believed to be crucial in developing girls into well-rounded members of society, healthy, fit and academically driven, though both Buss and her successor Sophie Bryant shared a concern that excessive interest in games would be to the detriment of health and schoolwork.[94]

As girls moved increasingly beyond the boundaries of the home and domestic roles, headmistresses and school medical officers alike had to gauge their positions carefully. They recognised the important role that parents would continue to play in caring for the health of their daughters and sought to engage them in supporting their health needs. In particular, homework and exercise were to be regulated at home, and mothers were expected to be present when their daughters underwent medical inspections. At least one father expressed his enthusiasm about the NLCS's regime; commenting on the introduction of musical gymnastics, he considered 'that his daughter's health has been greatly benefited by it, and that its use counteracted any ill effect the great amount of mental work, caused by her preparation for the Cambridge examination, might have induced'.[95] Dorothea Beale, in contrast to the approach adopted at the NLCS, declined to introduce medical inspections at Cheltenham, believing that the health of her pupils was ultimately the responsibility of parents. However, she encouraged parents to be more communicative on the subject of their daughters' health. The situation for boarders was straightforward enough in that the school took responsibility for the health of its pupils in terms of the regulation of study, sleep, exercise and diet. Day pupils were more problematic and Beale advocated regular meetings with parents, and insisted upon the girls having eight hours sleep, never working late and eating a good breakfast, and she sought to abolish stays and high-heeled footwear. She also urged mothers to obtain 'some *systematic* knowledge of the laws of health' via lectures, which they could attend at the school, or by reading relevant literature, and 'if they are not able to do this, they should conform as far as possible to our boarding-house regulations'.[96]

Even before the School Medical Service introduced medical inspection after 1908, making 'a vital contribution to the improvement of child health by alerting parents to the presence of disease and encouraging them to obtain medical treatment', medical men supported a role for schools in improving health in the home, or at least countering its negative impact.[97] Dr Howard Kelly explained that the helplessness and dependence engendered at home was a 'distinct hygienic disadvantage' for girls and the girl at school age was the product of her home environment. Girls' poor physical and moral training resulted from the 'unnecessary and thoughtless demands made upon the growing girl in poorly organized households', including long hours of piano practice, which interfered with the time available for open air exercise.[98] Schools, by means of medical inspections and the observation of girls in the gymnasium, were in a position to detect particular conditions, particularly scoliosis or curvature of the spine, which, Kelly pointed out – linking the benefits of schooling with future benefits for maternity – was vital because of the impact of this condition on the pelvis and mechanics of parturition.[99]

William Playfair suggested that one of the chief reasons for the more frequent breakdown of girls at school compared with boys was that 'the male's work is safeguarded by an amount of physical exertion in the way of sports which tends to keep him in health'; this was usually compulsory in boys' schools and optional for girls.[100] At the same time, childhood and early youth were seen as '*the* time when the surgeon, physician, or physical exercise instructor can obtain the best result of which the given child is capable'.[101] Though varying in the extent to which they introduced physical activities, by the late nineteenth century many girls' schools ensured that their pupils made regular appearances in the gymnasium, swimming pool and on the playing fields, stressing the positive effects of achieving a balance between mental and physical activity. Cheltenham Ladies College was set up in 1853 to provide the daughters of noblemen and gentlemen with an education that was adapted rather than an imitation of boys' training, and 'to furnish girls with a sound and balanced religious, intellectual and physical training in a manner that would magnify rather than threaten their womanhood', and its approach to physical training was notably less robust than that of the NLCS.[102] By 1890, however, a teacher trained in Swedish gymnastics had been appointed and a gymnasium fitted out, and a variety of sports were steadily introduced, though headmistress Dorothea Beale discouraged competitive games.[103] She considered hockey rough and one account claimed that she advised using more than one hockey ball to avoid over-exuberant play and injury.[104] Though, like Buss, an advocate of universal suffrage and education reform, Beale, under the direction of the schools' governing body, held more modest ambitions for her girls, and was 'most anxious that our girls should not over-exert themselves, or become absorbed in athletic rivalries'.[105] While dismissive of the value of 'accomplishments', the school syllabus catered largely for well-to-do girls destined for marriage and Beale assumed such girls would not have to earn a living. She was also opposed to competitive examinations, although able girls were allowed to sit them.[106]

Increasingly, approaches to physical training were debated and promoted by the Headmistresses Association, set up by Frances Buss and Dorothea Beale in 1874, and the Association of Assistant Mistresses, established in 1884. Their conferences engaged with such topics as hygiene, exercise, games, training in domestic subjects, infectious diseases and sanitation in schools, as well as the health benefits of holidays for school teachers. The associations noted the shift from the pallid Victorian girl, physically feeble, insipid and weak minded; 'in time', it was remarked at the Annual Meeting in 1888, 'we may hope to get rid of the tight-lacing, novel-reading, idle invalided young ladies, who bring so much discredit upon their sex'.[107] A report of 1895 asserted that 'the best answer to those who fear that improvement in education may be counter-balanced by deterioration of health, is the undoubted fact that the physique of girls and women has greatly improved under the new conditions of school life, in which the importance of outdoor games is fully recognized'.[108]

In 1898 *Work and Play in Girls' Schools* attempted to sum up the position physical training had taken in girls' schools by the turn of the century. The dominant message was the importance of a rounded education. Jane Frances Dove, headmistress of Wycombe Abbey School and a Girton graduate, produced the section on health and the cultivation of the body, declaring: 'The most important conditions for health are first of all a wholesome environment; secondly, wholesome occupation for the mind; and thirdly, proper exercise for the body.'[109] She advocated many forms of exercise – riding, rowing, bicycling and active games, tennis, fives, bowls, croquet, golf, swimming, skating, archery, rounders, tobogganing and basketball. Games were to be as 'joyous and spontaneous as possible', were 'essential to a healthy existence' and 'for people who are to be intellectual workers, games are the modern adaptation of the old command "to till the ground" '.[110] Dove outlined the special need for energetic games for girls:

> The joyousness and spontaneity are so especially necessary for girls, on account of their extreme conscientiousness and devotion to duty. Boys, for all I know to the contrary, may perform their duties equally well, but they are rarely inclined to worry over them as girls do, and they have such overflowing animal spirits that they always contrive to find relaxation, by means of fun and activity of all kinds at odd times, which does not come naturally to girls, or which if indulged in by them as well as by boys, would make life an unbearable pandemonium for their elders.[111]

Gymnastic exercise, according to Dove, was a vital component of school regimes, particularly Swedish gymnastics, which developed the girls' muscular system in a harmonious way. She warned against overdoing physical exercise, and for girls this was particularly important as so often their development had been neglected in childhood, but she also asserted that games promoted friendliness and cooperation, self-control, persistence and judgement. Chapter 3 outlined how increasingly sport for girls was described as character-building, encouraging cooperation and team spirit, and this was particularly pertinent, as McCrone has shown, in the context of girls' schools. Florence Gadesden, principal of Blackheath High School, with its reputation for excellence in athletics and games, emphasised that sport built character as well as mental and bodily health, allowing girls to train 'to manage [them]selves and others' and teaching them 'the necessity of fair play, give and take, courtesy to opponents, endurance of hard knocks, and so on – things which in those days were a novel part of a girl's training'.[112]

In addition to her discussion of sport and exercise, Dove emphasised the duty of schools in providing a suitable environment for study and other forms of activity, in encouraging good eating habits, and guarding against the too early development of 'emotions'.[113] Dove provided sample timetables for girls' schools, advocating that girls of 16 should only work up to six hours, 'the utmost a girl of any age ought to attempt'; overstrain could make girls 'anaemic and weak-backed, hating the sight of a book'.[114] Beyond exercise, sports and games, girls' schools evolved more complete programmes of monitoring and training in health that included teaching physiology, the laws of health and hygiene, and domestic science, as well as pastoral care. Dorothea Beale carefully monitored the health of her pupils, watching out for over-pressure or reports of home cares. In such cases, she advocated outdoor exercise, milk in the morning, a half-hour rest in the middle of the day, and early to bed, 'done in such a way as to brace rather than enervate'.[115] Sara Burstall, Headmistress of Manchester High School for Girls, described how girls' schools encouraged neatness, simplicity and good sense, and loose-fitting clothing; gradually, specialised dress was introduced for sports and gymnastics.[116] Even the journey to and from school across industrial Manchester was considered for its potential impact on health.[117]

The domestic arts or, domestic 'science' as it was becoming increasingly known, formed a significant part of the school curricula for girls in the latter part of the nineteenth century, something Sara Burstall believed to be vital given the encroachment of the school on girls' time to learn these skills at home. It was part of the commitment of schools in offering a broad curriculum and provided opportunities to less academic girls, as a wider range of pupils entered girls' schools, such as Manchester High School, by the early twentieth century.[118] Domestic science was perceived to be of national importance for girls of all classes, particularly given the concerns about national degeneration and high levels of infant mortality that peaked around this period.[119] With regard to this, domestic science did double duty; it promoted knowledge of hygiene, household management, infant care and food preparation and it saved girls from too much concentrated schoolwork. In 1908 Janet Campbell, Chief Medical Advisor to the Board of Education and a prominent infant welfare reformer, suggested that girls needed to be protected from too much schoolwork at puberty, especially lessons 'using up a great deal of brain energy', such as mathematics, and recommended as a substitute cookery, embroidery or handicrafts, which caused 'little mental strain'.[120] Burstall advocated 'the teaching of cookery and domestic arts to girls on national grounds, as is the teaching of military drill and marksmanship to boys'.[121] An Association of Teachers of Domestic Science was established in 1896, and by 1911 its members were arguing that 'a course of housecraft should form an essential part of a woman's education, and that all girls in Secondary Schools should be given the opportunity of taking such a course'.[122] Elizabeth Sloan Chesser's *Physiology and Hygiene for Girls' Schools and Colleges*, published in 1914, exemplified the ways in which teaching in this field had become more specialised and technical, instructing girls in science as well as good health, with its detailed explanations of dietetics and personal, household and public hygiene.[123] However, as well as ramping up the level of technical information, the book focused on mundane, practical household skills, and Chesser urged that girls should be provided with 'a comprehensive training in what will fit them to be efficient in the home sphere', including basic cookery, sick nursing and child care.[124] As Michelle Smith has suggested, 'the moral responsibility of household management was also significant in itself, imbuing the formation of curricula in girls' schools with importance for the physical and moral strength of the nation'.[125]

The pendulum swings back

Debates about girls' capacity to study without detriment rumbled on well into the twentieth century. The 'fixed fund of energy theory', as elucidated by Maudsley, proved durable, though far from uniformly influential. Headmistresses tended to override it, arguing that healthful exercise improved levels of energy and the physique of girls, and thus their ability to meet the challenges of the curriculum and even to improve performance across the board, though many emphasised the importance of establishing good health and bodily condition before turning to the intellectual demands of the curriculum. A number of second-generation headmistresses were themselves athletes, such as Lillian Faithfull, Dorothea Beale's successor at Cheltenham, who was an all-round athlete and President of the All England Women's Hockey Association.[126]

However, by the early twentieth century headmistresses were warning increasingly of the dangers of overdoing sport. It was suggested that too much sport put at risk female bodies at an important stage of development, while fierce competitiveness as well as the physical coarseness resulting from sport led to a loss of femininity. An article appearing in *Woman at Home* in 1912, with commentary from many prominent headmistresses, underlined many of these concerns. Lillian Faithfull at Cheltenham encouraged all forms of athletic exercise for girls, which she declared good for physique, morals and character, but warned that games should not be played to excess. Roedean was renowned for its sporting ambition and prowess of its pupils, and its three principals, Penelope, Dorothy and Millicent Lawrence, were convinced of the mutual benefits of a healthy body and mind, yet they also cautioned 'a certain moderation'.[127] Miss Leahy, headmistress of Croydon Girls' School, preferred the modern over-athletic girl to the 'neurotic and self-centred type of girl who seems to have been a well-known product in early- and mid-Victorian days', but nonetheless urged the 'moderate and sensible use of games and athletic pursuits'.[128] Sara Burstall warned of excesses in physical training and the 'intellectual and moral dangers of games and gymnastics among girls. They have only a certain amount of available energy.'[129] In her preface to *The Medical Inspection of Girls in Secondary Schools*, produced by the schools' Medical Inspector Dr Catherine Chisholm, Burstall pointed to the continued challenges the school faced in terms of the poor health of many of its pupils and the need for medical inspection and school hygiene to address these challenges effectively; she also commented on the increasing stress of the examination system, 'while the national ... demand for efficiency is greater'.[130] She referred to the work of eugenicist Dr Mary Scharlieb on the recreational activities of adolescent girls, who had urged the necessity of physical examinations ideally by a woman doctor ('there is nothing like setting a thief to catch a thief') before girls embarked on programmes of athletics, gymnastics and dancing. This would also enable the identification of the poor health of girls, their weak eyes, curvature of the spine, headaches and anaemia.[131] Scharlieb had also expressed concern that excessive athletics was producing the ' "neuter" type of girl' with 'a corresponding failure of function.'[132] Eugenicist physician Elizabeth Sloan Chesser strongly advocated out-of-doors exercise for girls, but also warned of the dangers of over-fatigue. Her concern, however – at least as expressed in this particular publication – was about the impact of fatigue on schoolwork rather than future motherhood: 'The girl who expends a great deal of energy in games at this transition time in her life is affecting to some extent her power of accomplishing intellectual work.' 'No girl can afford to over-tire herself when she has important examinations in view.'[133]

Miss C. Cowdroy, Headmistress of Crouch End High School and College, reflected back on her advocacy of physical culture for girls in a piece published in the *Lancet* in 1921. She also described how she had become increasingly disillusioned about the impact of drill and girls' take up of what had previously being considered boys' games. The girls became, she noted, 'more selfish, more concentrated on material things' and less womanly. Young women who joined in strenuous games 'became possessed of hard muscles, a set jaw, flat chest, and often a hard aggressive manner and an ungainly carriage'. They also suffered from nerves (manifested in irritable tempers), heart trouble, rheumatism, they ceased to menstruate, and their marriages were often childless.[134] Her publication prompted a flurry of correspondence in the press and the *Lancet*; her observation on the inability of athletic girls to bear children was declared of 'no value' and Dr Margaret Thackrah, medical officer at Dartford College, described students at physical training colleges as 'the most natural and normal girls that could be desired'.[135] A year later a Report on the Physical Education of Girls, produced by a joint committee drawn from the Royal Colleges of Physicians and Surgeons, the Medical Women's Federation, the Ling Association and various organisations of teachers and based on surveys taken from medical practitioners, female medical students and headmistress of both state and independent schools, expressed the majority view that games and exercise were beneficial to character. Only a small number of contributors thought that their value was overemphasised. Apart from physical fitness, games and sports were reported to 'conduce to alertness, resourcefulness, and judgment, and encourage a public-spirited and healthy outlook'. The report also concluded that there was no 'clear proof' that strenuous physical exertion had any influence on the prospect of motherhood or difficulty of labour, though a minority again affirmed that it was detrimental to home and other interests and lessened 'womanly qualities'.[136]

The report also debated the restriction of physical exercise during the menstrual period, arguing that evidence increasingly suggested that this was harmful rather than beneficial.[137] Prefacing the views of the Medical Women's Federation, Catherine Chisholm emphasised that 'Many erroneous ideas are held on the subject of the health of girls in adolescence... the onset of menstruation ought to make little if any difference to her daily life', though girls suffering pain or profuse bleeding required medical attention and cases of debility or anaemia needed rest and treatment.[138] Chesser cautioned girls to take care during menstruation, and particularly to avoid catching chills or becoming over-fatigued, but 'most girls do not experience any adverse symptoms during this period... and there was no reason for any girl of normal health to alter the day's routine or to give up her usual work or recreation'.[139] The Medical Women's Federation, as part of their campaign to encourage a rational approach to menarche, provided authoritative advice on the subject, including information intended for schoolgirls, advocating the continuation of ordinary work and play, games and gymnastics during menstruation.[140] Dr Christine Murrell advised in her comprehensive volume *Womanhood and Health*, which featured sections devoted specifically to adolescence, that the body should be kept scrupulously clean during menstruation and was keen to dispel 'old wives' tales' on the subject.[141] Dr Alice Sanderson Clow, one of the Federation's most active members, concluded on the basis of her research on 1,200 pupils at Cheltenham Girl's School where she was Medical Inspector, that most school girls were free from menstrual disturbance, that girls should be urged to take baths and to exercise during their periods, and that 'study *per se* is not a cause of dysmenorrhoea, although if pursued to the exclusion of daily exercise it may be indirectly a contributory factor'.[142]

This dovetailing of concerns about the impact of games, de-feminisation and the risks for future motherhood in the early years of the twentieth century drew on a heightened interest in motherhood and the influence of eugenics discourse more generally, as well as broader anxieties about the challenges of youth that were explored in Chapter 1. But, just as the views of the medical profession on the relationship between girls' health and schooling varied in the late nineteenth century, during the early twentieth century there was again much diversity of opinion amongst eugenicist authors. Miss M.E. Findlay's grim account, for example, echoed Maudsley's assertion that energy supplies were fixed, and that 'taxing the brain at a time when the reproductive organs are acquiring their mature form, is... a direct cause of sterility'. In early adolescence girls worn down by the 'disorganisation of the monthly function', Findlay argued, should be withdrawn from games and athletic displays, and between the ages of 14 to 16, or possibly 15 to 17, school hours and school tasks should be 'distinctly decreased, and the time thus freed devoted to home duties'.[143] Mary Scharlieb, conversely, while expressing reservations about the extent of girls' engagement in sporting activities, overall found them invaluable for their prophylactic, remedial and moral impact, equipping girls with self-respect and good temper; games enabled the woman 'to take her part in the serious business of life', in the home, civic or professional spheres.[144] Elizabeth Sloan Chesser, meanwhile, framed 'education' in the broadest terms, and declared that all girls should go in for 'higher' education of a type which would fit individuals to be useful, healthy and efficient.[145] While urging girls to learn how to manage their energy resources, she encouraged them to be ambitious:

> A generation ago very few girls thought of asking themselves, at fifteen or sixteen years of age, what they were going to do with their lives. The idea of a girl following any business or profession was undreamed of, but it is becoming more and more usual for young girls to decide upon some special line of work, just as if they were boys... Let it be your ambition to be an all-round girl, good at games and good at lessons, able to cook a dinner or make a speech.[146]

Conclusion

Thus we have a situation where some headmistresses expressed reservations about girls' tendency towards overexertion, while some eugenicist physicians advocated the benefits of training in sport for the all-round development of girls, placing an ambitious set of targets before them, to be prefaced by the establishment of good health and good health practices. Anxieties about racial and national degeneration had penetrated deeply by the turn of the century, and many headmistresses echoed these broader anxieties, and pondered too the wide range of demands being made of adolescent girls. By the turn of the century very different medical perspectives established themselves amongst those engaged with the relationship between girls' education and health. These ranged from those of Clement Dukes and a growing body of female school medical inspectors, who had direct contact with school children and educationalists in their daily practices and worked with schools as potential sites of health promotion, to psychiatrists, such as Crichton-Browne, steered strongly by what remained powerful medical theories and confronted with the apparent effects of over-study in his psychiatric practice. Dr Robert Jones, Medical Superintendent of London's vast Claybury Asylum, quarrelled too with 'the educationalists of to-day, and I think high medical opinion is with me' in driving girls from classroom to sports field without pause or rest, making 'the body a motor machine' and causing 'the "insurgency" of the modern girl against domestic responsibilities and the common duties of life'.[147] The lack of coordination between mind and body the system enforced, 'the wild cult of athleticism' and the 'frantic rush' led, according to Jones, to the worse kind of nerves or neurasthenia, as well as to the 'cricket stoop', the 'hockey walk' and the 'golf stride'.[148]

For headmistresses, concerns about the health of their charges persisted but were rather different, preoccupied as many were, not with 'nerves', which they argued had been largely abolished, but by the persistence of poor physique and very obvious signs of defective growth, and this was exacerbated when girls' high schools, an expanding sector, began to recruit their pupils from broader social backgrounds. In such situations, 'Gymnastic exercises, wisely planned and executed' were promoted as the key to 'producing a fine and proportionate physique'.[149] Schools strove overall to develop a wide-ranging approach to health and in this sense functioned as much as sites of good health practice as centres of feminist emancipation, and indeed many schools began to place increasing emphasis on domestic skills. Meanwhile, as Fletcher has suggested, there was never a better time to sell physical education – and indeed ideas on health more broadly – to a middle class interested more generally in the widespread uptake of sport and health matters.[150] By 1900 *The Girl's Realm Annual* could proclaim that 'the finest healthiest sites in the county are chosen for the building of the modern school', sport was lauded, and 'girls grow up well-informed, alert, and beyond everything healthy'.[151]

Also at stake was the recruitment of 'the girl's own earnest cooperation' in striving for perfect health and overcoming physical difficulties and rare testimonies from the girls themselves emphasise that they thought about health and enjoyed the benefits of the mixed curriculum.[152] An article written by a group of girls for the NLCS *Magazine* in 1885 commented on their pride in the school's gymnasium, with its parallel and horizontal bars, ladders and other paraphernalia, as well as the need to engage seriously with gymnastic work and to improve their physical status. Reflecting back on their predecessors in earlier decades, the article described how girls who formerly did 'their

Calisthenic exercises in a listless, lazy style, are now compelled to do good honest work, whether they like it or not. In short, it is hoped that before long there will not be one round pair of shoulders, or stooping back to be seen in the School.'[153] A decade later a 'Girl Graduate' described life at our great public schools in the periodical *Health* in overwhelmingly positive terms, explaining how the girls' workload was carefully planned, and would involve six or seven hours study a day. There was no 'rush' of work; 'the lecturers are most considerate, and are usually inclined to check, rather than to urge to an increased amount of work... Six or seven hours of intellectual effort are not too much for any.'[154] She went on to assert 'that strenuous exercise of any faculty is distinctly beneficial, as raising the "tone" of mind and body'. Girls spent time in the fresh air and many were first-rate at tennis, hockey, golf, cricket and fives, 'and frequently stand two to three hours a day in such games, which completely take the mind off all work, take the girls out-of-doors, and provide a thoroughly healthy mental discipline and physical training, that turns out some of the most superbly healthy and well-developed girls I know'.[155]

5
The Health of the Factory Girl

> Their frivolous air excites our pity; the extreme youth of many
> makes us tremble to think of their perils in going to and
> from their work, ... and we also remember that in business
> they may be thrown into the company of careless, worldly
> men .... In some cases the mother is indifferent and easy-going,
> and quite content that her child should begin to earn some-
> thing ... When the factory closes, home not being a bright or
> inviting place, the weary girl craves some change, some amuse-
> ment, and too often hurries off with a companion to a music
> hall or dancing saloon.[1]

This emotive language was employed in 1885 to draw attention to
one of the principal aims of the Young Women's Christian Association
(YWCA): to care for the moral wellbeing of working girls, as they faced
the temptations of the city, moving through an urban landscape full
of danger to their places of work, lured by the bright lights and inappropriate pleasures away from home, itself unappealing and indifferent.
The purpose of the YWCA was to rescue and befriend such girls, to
domesticate them, lead them away from the ruin and perils of the city,
and draw them towards Christianity. Additionally, the YWCA rapidly
absorbed itself in activities directed towards improving the health of
young women, an aspect of their role that has received scant attention.
Such activities led the organisation to frame its work not only as being
directed towards inspiring Christianity, moral rectitude, good behaviour
and respectability, but also to take account of the challenging environmental conditions which confronted young women workers on a daily
basis. Indeed, the above quotation was in many ways as much about
environment and health as temptation and corruption. Health itself was

broadly, even imaginatively, framed by the YWCA and the other girls' clubs to be explored in this chapter, encompassing conditions in the workplace, the home environment, movement to and from work, recreation, exercise and rest, diet, hygiene and body care, and spiritual as well as physical wellbeing.

The working and home environments of girls as well as their experiences of schooling have been widely explored for the late nineteenth and early twentieth centuries, and so too has the socialisation of young working women and their leisure time.[2] June Purvis's model of the 'double burden' of class and gender, which, she argued, operated to limit the opportunities available to young girls, provides a useful framework for considering the challenges young working women faced during this period.[3] Though for some opportunities no doubt expanded and their earning power increased, many girls would continue to face another form of double burden as, following long working days, they came home to heavy domestic duties. While it has been widely recognised that working-class girls in these situations must have been exhausted and run down, their health experiences have received relatively little consideration. Irvine Loudon, however, has described the high rates of anaemia and general ill health amongst working-class girls and domestic servants which preoccupied doctors working in large city hospitals and dispensaries by the end of the nineteenth century, while Anna Davin has pointed to the poor health of girls brought up in poverty, including the 'perennial' problem of headlice, endemic in poor, overcrowded districts.[4] Meanwhile, a number of studies have explored the health challenges faced by girls employed in industrial labour, notably in the First World War, and their liability to industrial diseases, poisoning and accidents.[5]

Interrogating the archives and printed resources of the YWCA and other girls' clubs, this chapter focuses on efforts to improve the health of young women, as they came to comprise an increasingly large and visible component of the workforce. It examines how the health of working-class girls was represented and how health messages were conveyed via the journals and gazettes produced by these organisations and in pamphlet and prescriptive literature. Focusing on the 1880s through to the end of the First World War, it will be suggested that, while the war presented a special set of circumstances, the emphasis across this entire period in terms of girls' health lay, not with their biological limitations, but on the influence of their behaviour, as well as the environments in which they found themselves.

Historians examining the industrial work of women in Britain in the early decades of the twentieth century have tended to emphasise the dangers such work was believed to pose to women or girls as 'mothers (or future mothers) of the Empire'. Barbara Harrison has argued that the primary objective of state intervention was to protect reproductive health; consequently, married women were targeted despite the predominance of young unmarried women in the workforce.[6] Similarly, Carolyn Malone has characterised the legislative restriction of women's work, implemented through the Dangerous Trades legislation, as a policy of foetal protection which sought chiefly to preserve women's reproductive health.[7] Harrison and Malone's research supports a large historiography which asserts that women's health was chiefly taken up by the state through the development of maternity and infant welfare policies.[8] Yet investigation of government reports and the archives and journals of agencies interested in industrial health has indicated that there was a great deal of interest in the health of young, unmarried girls in the workplace, and much of this interest addressed a wider range of issues than future reproductive capacity, though this certainly gained additional resonance in wartime.[9] As with middle-class girls, the promotion of positive health for working girls focused on girlhood as a distinct phase between childhood and marriage and motherhood, though the content of advice on this subject – and the opportunities for achieving health – were very different. Welfare workers, club leaders and factory inspectors urged improved hygienic practices, attention to diet and access to healthy activities, and consideration of girls' psychological health was accorded an increasingly important role. Much of this activity was not directed at improving girls' status as future wives and mothers, but rather aimed at fulfilling more immediate health goals.

Working girls and their environment

Girls were typically low down in the pecking order in terms of the building blocks leading to good health. Often described as undernourished – indeed the 'factory girl' was typified in contemporary accounts by her small and scrawny frame – if they worked they were liable to fare better than their mothers, but usually worse than their fathers and brothers.[10] With regard to food, mothers, according to Robert Roberts, felt that their needs were 'not the same as lads' and none looked 'more pathetically "clemmed"' than the little schoolgirl'.[11] A long working day in the factory or domestic service was often followed by heavy household work and laundry or the care of younger siblings, with little scope for recreation and exercise. The housing of working-class girls was typically overcrowded and lacked labour-saving devices. 'Mothers generally expected more of their daughters than of their sons ... girls' tasks, like those of their mothers, were never finished.'[12] Those that worked away from home were forced to opt for poor lodging houses, where they were unwelcome during the day. Those that remained at home could be the sole breadwinner, which one female Factory Inspector pointed out in 1906 could lead to girls working excessively long hours. Miss Vines cited one case of a girl getting up at 6am and returning home at 11.30 pm. Another, 'A girl of 16, in a bookbinding factory, looking more fit for bed than work. "Yes," the foreman tells me, "her father is out of work, and her wages, varying from 6s. to 8s. a week, have to go to supporting the family; she does not get enough food; that is what is the matter with her." ' Vines also reported that problems involving overheated, poorly ventilated, overcrowded, damp and unsanitary workrooms were rife in factories employing young girls.[13] While Vicky Long has cited instances of industrial welfare provisions benefiting girls, 'it was more common...for factory inspectors to bemoan insalubrious conditions than to commend good provisions'.[14] Girls' prospects for achieving good health appeared to have been limited, while the work itself, if not actually damaging to health, was certainly unlikely to boost it.

After the establishment of the school medical service in 1908, girls underwent health inspections while attending elementary school and had access to a limited range of medical interventions to tackle poor vision, dental defects and ringworm.[15] The service also rapidly came to embody the idea that the State should not just impart knowledge on health to children 'but that it must also see that the child is capable of assimilating that knowledge, and that his environment is not such that it will entirely undo the effect of...school training'.[16] At a point when increasing numbers of girls entered the workforce and were regarded as being at risk of losing skills in homemaking, emphasis was placed on the necessity of schooling them in domestic economy and transforming them back into 'little mothers'. In practice, provision in schools was patchy, while teaching on health was scanty, non-existent, or irrelevant and of the simplest kind, such as, 'early to bed and early to rise' or 'keep yourself clean'.[17] Even domestic science lessons tended to be largely theoretical and one school inspector in 1877 criticised their irrelevance to daily life; rather than being instructed about albumen, fibrin and casein, the children should be taught, he argued, about the cleansing of drains, how to dispose of domestic waste and how to treat cuts, burns, colds and simple diseases.[18] As we have seen in Chapter 4, drill and other exercise regimes were, however, steadily introduced into girls' elementary schools, and the theme of continuing the work that schools had initiated was taken up enthusiastically by the YWCA and other clubs. Clubs were considered vital in imparting health messages, guiding physical development and providing structured recreation to girls seemingly lacking any kind of direction. With no access to the kinds of health interventions that middle-class girls enjoyed, including tailored inspections and regulated exercise at school, as well as recourse to family doctors, health advice for working-class girls was most likely to be centred around club activities after leaving school and, if they were fortunate enough to have a forward-thinking employer or an active trade union, the workplace.

So how much and in what ways did working girls gain access to the positive health messages that we have seen to be so pervasive by the late nineteenth century for middle-class girls? Certainly, as Lucinda McCray Beier has demonstrated for working-class Lancashire, as periodical publishers reached further down the social scale, their mothers at least were likely to have become avid magazine readers, particularly of penny domestic weeklies, which contained information on medical matters, as well as recipes, tips on housekeeping and childcare and aids to beauty.[19] Young women also had increasing access to magazines catering for factory girls, though, as Penny Tinkler has pointed out, the intended and actual readership of these papers were not always synonymous.[20] So-called mill-girl papers were produced in response to the increased spending power of working-class girls and included the *Girl's Best Friend* (1898–1931), *Girl's Weekly* (1912–22) and *Peg's Paper* (1919–40), which largely focused on romantic fiction and fashion and beauty tips.[21] Some may have purchased the *Girl's Own Paper* (*GOP*), which certainly expressed a desire to reach working-class readers, though, at the same time, it regarded girls forced to work for a living as rather strange anthropological subjects as well as appropriate objects for the sympathy of its predominantly middle-class readership. Certainly, the *GOP*'s health columnist, Dr Gordon Stables, regretted 'the plan so common in this country of putting girl-children to work so young'.[22] The *GOP*'s editor Charles Peters, meanwhile, wished to cater for 'girls of a less high position' who would receive instruction in 'economical cookery, plain needlework, home education and health'. Its content, however, aimed largely at middle- and lower-middle-class girls, and much of the material on appropriate employment for girls referred to careers as artists, writers, sanitary inspectors, doctors and teachers, rather than work in the factory or domestic labour. However, during the First World War the *GOP* included articles on munitions work and other heavy jobs filled by women, praising girl chimney sweeps, bricklayers and land girls, and emphasising the crucial contributions of young women to the war effort.[23]

Though concerned about girls' early uptake of heavy work, in 1885 Gordon Stables also described working girls as particularly sturdy and their energy an important form of capital. He also somewhat naively recommended them not to 'overtask' themselves at work and to 'pull up in time' if they observed feelings of 'lassitude or weariness', signs of insipient illness.[24] Another anonymous article appearing a year later painted a more miserable picture of the plight of working girls, often deprived of air, food and exercise, which further depleted their 'less robust frame' and 'smaller powers of endurance'.[25] A number of advice books included tips for working girls as part of their remit, and some, like Curwen and Herbert's *Simple Health Rules*, published in 1912, were specifically targeted at working girls and women, seamstresses, clerks and shop girls, prefacing a stepped-up concern to impart advice to girls working in industry during the emergency conditions of the First World War.[26] Other authors of advice literature, as well as advising girls on how to manage work without detriment to their health, also reported more generally on their plight:

> Girls are ready for work at an earlier age than boys, and have often to submit to severe muscular endurance at a very early age. The simple exertion of standing for hours together is nowhere better exemplified than in the class of shop girls. The number thus employed in this country are hundreds of thousands . . . and it happens that the physique of an enormous number of healthy girls is destroyed by this seemingly simple labour. In the first place, the fact of being indoors from morning till night is terribly against health; next standing at and leaning forwards over a counter is a severe muscular strain, especially on the muscles of the back and the calves of the legs.[27]

Sustained standing for weeks on end was likely, it was concluded, to result in varicose veins or leg ulcers.[28]

The health of working girls was depicted as being bound up closely not only with the challenges represented by the environments in which they lived and worked but also by their behaviour, and Vicky Long has pointed out that female Factory Inspectors were liable to put forward initiatives designed to improve girls' moral welfare as much as their health care.[29] Working girls were described as frivolous and careless in their attitude to health, spending their wages on high heels, fancy clothes and make-up, scoffing buns and drinking stewed tea rather than eating sensible meals, preferring to lurk on street corners or in dance halls instead of engaging in more fulfilling recreations. They were perceived in many ways as vulnerable and pitiful, but also as rough and unruly, as letters to the *Times* described them in 1901, 'Hooligans of the Female Sex'; 'Their lives... a mixture of vice and intolerable monotony'.[30] Such assertions were fuelled by the increased visibility of girls in the workplace, as the employment of young women soared in the late nineteenth century; by 1901 over half of all women workers were under 25 years of age.[31] At the end of World War I more than three million girls in Great Britain were employed in factories, workshops or in domestic service, and census figures showed that 44.8 per cent of girls aged 14 to 15 were in full-time employment by 1921.[32] In some areas of the country, the rates of employment of young women were much higher. In 1921, some 50.8 per cent of girls living in Barrow and aged between 15 and 19 worked full-time, a figure that rose to 86.7 per cent in Preston.[33] In 1910, Elizabeth Sloan Chesser commented how 'one of the first things that strikes a visitor to a factory is the large number of women and girls at work compared with men'. It was a 'depressing spectacle'.

> Most of them are languid, expressionless, anaemic; many are mere children, 'half-timers' of from 12 to 14, or 'young persons' of from 14 to 17 years of age. The younger girls are under-sized, sallow and anaemic; the older women bear the marks of excessive strain in their thin, worn faces... Both the older women and the girl workers look insufficiently fed; they look as if they never breathed fresh air; they look cheerless and sad.[34]

Girls' clubs, visions of health and responsibility for health

As the industrial landscape of late Victorian Britain became populated by working girls, they became targets of intervention by the YWCA and other girls' clubs. The YWCA in particular provides a comparatively early case study of activities intended to enhance the health and wellbeing of young working women, and also allows us to reflect on how far these adhered to or acknowledged dominant medical discourses of the period and the issue of girls' future role as mothers. Concerned not only, but largely, with working-class girls, the organisation also strongly promoted domestic work for young women, and its support extended to factory hands, shop girls, teachers, clerical workers and nurses. The YWCA and other girls' clubs tackled a broad range of health issues – hygiene, diet, exercise, recreation, fashion and beauty, mental health, morale and the impact of the boredom and monotony associated with industrial work – through their publications and club activities. Additionally, the YWCA and a number of other clubs expressed concern about protecting its girl members from moral waywardness and promoted total abstinence, ambitions which often dovetailed with health concerns, the two mutually reinforcing each other. Healthy girls were more likely, the YWCA asserted, to be pure in body and mind, and the preservation of health was presented as a Christian duty and responsibility.

The YWCA represented the fusion of two initiatives – a prayer union set up in 1855 by Emma Robarts and the establishment of a home for returning Crimea nurses by Mary Kinnaird in the same year, which branched out to provide a network of missionary homes in London. It evolved into a national organisation with multiple branches (some 130 by 1872) and later into an international organisation. Rapidly, one of its primary goals was identified as ensuring girls' moral and religious wellbeing as they entered the risky arena of the workplace, where they would be exposed to poor working and living conditions and low pay, while the pressure of overwork was described as likely to impair bodily and mental health. Sweated labour became another major preoccupation of the YWCA, alongside industrial accidents, illegal hours and the shortfall in the number of female factory inspectors.[35] The YWCA closely affiliated itself with the National Union of Women Workers and National Federation of Women Workers and supported female trade unionism.[36] Given the expansion of the female labour force by the late nineteenth century, the YWCA and other clubs had the potential to be influential forces. By 1924 the YWCA had 450 clubs used by around 40,000 girls each week.[37]

In terms of health matters, the YWCA had a broad and creative remit, and saw health as being closely integrated with other club activities, including education and recreation (see Figure 5.1). The organisation informed and advised members on a range of health issues largely via its monthly *Gazettes*, particularly general health and wellbeing, hygiene, food and nutrition, exercise and sports, training in sick nursing and baby care, and tips for dealing with straightforward medical complaints. A rigorous approach was adopted towards exercise and physicality, with early issues of *Our Own Gazette* urging the uptake of sports, including rowing, kayaking and canoeing, skating, cricket, field sports and races. Its very first issue bore the image of a young woman propelling herself forward in a kayak, emblematic of the way the organisation saw itself as projecting girls into a better future (though failing to consider perhaps how many readers would have access to such an activity).

Figure 5.1   New YWCA Institute Buildings, Liverpool.
Source: From *Our Own Gazette*, I: 11 (November 1884), pp. 126–7 (Modern Records Centre, University of Warwick, MSS.243/5/1: YWCA/Platform 51).

Literature on women and sport has tended to bypass club material, aligning the emergence of exercise, sport and games for young women with the rise of girls' schools and women's entry to universities. However, the YWCA directly facilitated a wide range of health-promoting activities for working girls through its clubs: drill and gymnastics, swimming, hockey, netball, rounders, and Greek and country dancing. As part of its broader mission to house and care for young female industrial workers, the organisation provided hostels, canteens, holiday homes and convalescent facilities. The YWCA set up a Health Section which encouraged girls to take up Health Insurance and a Social and Legislation Committee in 1918, which became the Industrial Labour Bureau in 1921. Aiming to tackle poor conditions and breaches of factory legislation in non-unionised workshops and factories, the Committee took up such issues as the poor provision of canteens, cold and unsanitary premises, and excessive hours of labour, as well as informing girls of their rights under industrial legislation.[38]

The loss of discipline and structure imposed at school, and the negative impact this had on health, was an enduring concern, as was the corrupting influence of the workplace and the problem of landladies who discouraged lodgers from staying in their rooms during the day. The Leamington Spa branch of the YWCA set up a temporary home for young women engaged in business in the 1870s, where evangelical work was carried out, reporting that over five years of those cared for in the home only 5 per cent had 'turned out badly'. One skilled embroidery worker had lost a good situation, and, threatened with starvation, was tempted by suicide. After support in the home, she gained a situation as a mother's help.[39] Factory, street and spending money represented a potent mix and a major health risk: 'On leaving work', declared one club correspondent describing the straw bonnet workers of Dunstable in 1880, 'they parade in town till 9.30, which is very bad for their health, as the workrooms are very warm and the air of Dunstable particularly keen'. The girls were declared to 'have a great opinion of themselves & are a difficult class to deal with'. In response a reading room was provided with the assistance of 'some nice young women belonging to different Bible classes in the town'.[40]

The YWCA's work in health promotion was interlaced with ambitions to improve the moral standards of its membership, there is little question of that. Yet it also configured efforts to improve girls' health as part of a broader aim of tackling poor conditions and providing good facilities for working girls, to house and feed them, and to offer them recreational outlets. Its concerns were with the environments within which girls often found themselves, often separated from parental influence – though this could be construed positively as well as negatively given the conditions which often prevailed in the girls' homes – in lodgings, travelling to work and in the workplace itself. The organisation strongly urged its members to abstain from drink, yet also recognised that alcohol abuse was related to worry and weariness, long hours of labour and rushed meals. While girls were perceived to be easily corruptible and subject to a range of foibles, amongst which a rashness with money and love of inappropriate finery featured strongly, as did a lack of regard for personal hygiene and a tendency towards laziness, they were seen as having a great deal to contend with in terms of living and working conditions.

The YWCA recognised that many working girls were under constant pressure to combine work and domestic roles; after a long hard day in the factory, they were liable to return to domestic tasks at home, whether this was the parental home or lodgings, where girls, ill-equipped in terms of cooking skills and facilities, had the task of preparing their own meals. The provision of hostels and canteens for working girls was one way of mitigating the problem for girls living away from home, and also helped wean them off diets described as being composed predominantly of tea, buns and pickles and on to properly cooked meals, accommodating them in homely and tightly controlled spaces, where a 'religious atmosphere' dominated. In making a plea for the provision of restaurants in 1885, Miss Trotter commented upon how

> The only way which these sisters of ours *can* economise is in their *food* – precisely the point where the economy is most dangerous. Toiling twelve hours a day in cramped position and close air, and experimenting on how little will sustain life, what wonder if one after another breaks down in health; or worse still, takes to stimulants to restore the 'sinking' produced by want of wholesome, nourishing food. 'Do give me one of your price lists for my sister,' said one of our customers the other day, 'She has made herself quite ill living on pastry... I was losing all my strength before *you* opened,' said another... I live at the other end of Regent Street, and the eating houses about there are so unpleasant – all are full of men. I have gone without dinner day after day rather than enter them.'[41]

While expressing concern about the limited resources available to girls in terms of access to nutritious food, the YWCA was also perturbed about declining interest in domestic matters amongst working girls which they claimed was one of the most pernicious influences of the factory and workshop. Emily Kinnaird complained that all industries open to women 'are unfitting them to become good wives and mothers'. She proposed that the YWCA must *'meet this growing evil'* by setting up domestic science departments teaching cookery, nursing and dressmaking, and advocated the use of cheap books to educate country girls.[42] In some respects the YWCA *Gazette* adopted, the 'familiar, cozy tone' which Lucinda McCray Beier has suggested marked out women's magazines, with its authors depicting themselves as familiar with the settings within which girls found themselves.[43] The *Gazette* appears, however, to have been unlikely to accept its readers' customary ways of doing things and aimed to change behaviour and practices. In 1900 it exhorted girls to eat properly, particularly girls living alone, and provided recipes designed especially for single girls – scrambled eggs, fish, cheap meat dishes and stewed fruit – tips for keeping food fresh, and on budgeting, and insisted, 'Never, however tired, sit down to a soiled tablecloth or serviette and *never* used dimmed and greasy spoons and forks.'[44] A feature appearing in the same year explained the 'evolution of a dinner' from a stale loaf.[45] Yet, in emphasising proper meals and their proper presentation, the YWCA was keen to enable girls to improve their own nutritional standards as well as their skills as future homemakers. Setting up orderly canteens offering healthy food at reasonable prices also became a cornerstone of club work. The Manchester Girls' Institute in Ancoats organised a restaurant offering low-price meals to around 200 girls every day, while the Emily Harris Home Dining Rooms, which advertised in *Girls' Club News*, offered affordable food to working girls, with hot meat and vegetables costing 4d and pudding or fruit salad 1d.[46]

In explaining girls' poor health and the challenges that they faced in improving it, the impact of both environment and behaviour were invoked again and again in the *Gazettes* and annual reviews of the organisation. Inappropriate behaviour often trumped the influence of their surroundings. Girls were depicted, for example, as letting themselves down in terms of their hygienic practices, which resulted partly from a lack of knowledge and encouragement on this score, but also from a tendency towards laziness and an urge to mask deficits in the care of skin, hair and bodily hygiene with artificial aids. Hygiene was equated with good health and beauty seen as a product of robust health and cleanliness rather than cosmetics, and the YWCA warned of 14- to 18-year-olds who quickly acquired the superficial appearance and manners of adults, 'with lipstick and high heels'.[47] Tight-lacing, one *Gazette* article declared, was 'morally wrong' as well as injurious to health, 'an actual sin against Nature, if we do not strive to at least help the bodies given to us in a sound and healthy condition'. The article went on to describe girls aiming to achieve waists of 19 to 21 inches, and cited the case of a girl dying of consumption, whose doctor attributed this to tight-lacing. 'Girls', the article concluded, 'have no right to make themselves so delicate that as soon as responsible for a family she becomes a confirmed invalid, besides the wrong of transmitting feeble constitutions to our children'.[48]

Dental care was a persistent area of interest, and *Our Own Gazette* urged girls to brush their teeth twice a day, to floss with a quill toothpick, and to visit a dentist every three to six months, commenting that 'Many girls to-day who have lost their own teeth could have retained them had they been less lazy.'[49] Similar concerns were shared by other girls' organisations. The *Girl Guides' Gazette* described the mouth as 'The Gateway to the Body' and neglect of the teeth could lead not only to tooth decay but disease, including enlarged tonsils, sore throats, discharging adenoids, anaemia, indigestion, diseases of the lungs, tuberculosis and pneumonia, as well as to a bad complexion. 'Not only is a neglected mouth a danger to the owner's own health but it also a danger to the community. With every cough or sneeze millions of disease germs are scattered about.'[50]

The YWCA directed its attention to the promotion of physical culture and exercise at a relatively early stage. In 1884 A. Alexander, director of Liverpool gymnasium, published a piece in the *Gazette* advocating gymnastic exercises for shop and factory girls, particularly for those whose 'sedentary occupations induce an apathetic manner and a stooping gate [*sic*] and posture', the forerunners of consumption and other related disorders and recommended firms as well as YWCA committees to support this work.[51] The listlessness of girls who hung around gossiping idly could be countered by sport and exercise, and, in an article on 'Healthful exercises for Girls' which also appeared in 1884, Miss Kate Alden declared:

> I know, from practical experience, that such exercise as I have described, although perhaps tiring, is yet eminently conducive to a healthy life; and I am quite sure that if some of our elder girls, who complain of lassitude and want of appetite, were to scull a mile or two against stream, or take a good six or seven miles' walk, they would find a cure for many of these evils.[52]

Girls were enjoined to take small steps towards improving their fitness – getting off the bus or tram one stop before work saved money (which was then ideally to be deposited in a charity box), and involved a brisk, refreshing walk – 'what a difference it will make to your health!' Skipping was urged for the stout, while thin girls 'should still walk; walking will give you such an appetite that you will digest your food more easily, and so improve your general health', while deep breathing would ensure that oxygen carried about the body resulted in increased vitality. 'Bright eyes, clear skin, rose-pink cheeks, are some of the outward effects; and tranquillity of temper and an absence of nerves result from the fact that the whole nervous system is receiving oxygen in full amount.'[53] In a series of lectures on 'Health and Hygiene', published in the *Gazette* in 1920 the benefits of long walks in good boots was suggested to promote health and godliness in equal measure: 'Hold yourself upright, keep your head up, thank God for life.'[54] Girls were to strive for a positive frame of mind and physical and spiritual cleanliness. 'Don't think about being ill, don't talk about diseases. When you get up in the morning, realise that you are at the beginning of a new day, God has got some work for you to do, you have got to be fit and well to do it.' 'Be very proud of your body and take every care of it. Let us remember that it may be the Temple of the Spirit of God Himself.'[55]

Other *Gazette* writers suggested more rigorous regimes based on gymnastic exercises or 'sporting stunts'. In 1893 marching, deep breathing, stretching and equilibrium exercises were advocated for the sick and convalescent. Girls were urged to use special apparatus to overcome defects, such as weak lungs, curved spines, uneven hips and shoulders – 'To be able to lift the weight of her own body, is one of the conditions of a perfect physique in a woman.' Walking was advocated in the early morning or after work, for distances of up to six miles, which would banish pallor, strengthen the lungs and improve appetite. Gymnastics helped make women dextrous and nimble, to forget how she looks, and 'requires some forgetfulness of that innate timidity and dread of criticism'; it 'prepares her for great or trifling emergencies, gives her an exhilaration of spirit, and a pleasure, which makes the exercise a delight and enables her to do more work with less weariness than before'. All exercises, the article went on, were to be followed by a cold bath and vigorous rub.[56] S.F.A. Caulfeild, explained that clubs' encouragement of physical improvement would also by 'a reflex action produce a healthier condition of the brain'.

> The brain cannot be highly cultivated and actively employed, and the rest of the frame left in comparative inertion [*sic*]. A fair balance of work and of rest should be kept up throughout the whole system. Thus the existence of clubs to promote the bodily powers is highly to be approved... Many girls belong to that sterling class known as 'muscular Christians:' and if they be not tempted into the grave error of supposing that their more delicate frame and general organization is designed for an equal amount of physical exertion as their brothers... their 'muscular Christianity' will serve them well.[57]

The Girls' Friendly Society (GFS), set up in 1875 by Mrs Mary Townsend in response to her concerns about unmarried girls leaving the countryside to work in large towns as servants or as factory workers, promoted a variety of physical exercises, country dancing and organised games, as well as nature walks, though these were not to 'degenerate into an uncontrolled romp'.[58] The Girls' Brigade, with its stronger military bent, was lauded, meanwhile, in the *Girl's Realm* in September 1908, for spreading the privileges of drill and training to the 'weakly and miserable' urban poor of London:

> Watch a poor, wizened little wisp of a girl for a week or two after she has started drills. You will see her developing a grace you would

not have suspected. The clubs and dumb-bells make her limbs firm, supple, and pliant; the breathing exercises bring her chest and shoulders into shape; the walking exercise teaches her how to carry herself. The natural result of this is increased cheerfulness, for physical health imparts its benefits to the mind.[59]

When exploring the health of working girls we might want to add to the double burden of class and gender identified by Purvis the further challenge of 'youth', which was represented increasingly around the turn of the twentieth century as a period of intense risk to the future development of both boys and girls by those working most closely with them. This was shaped, as seen in Chapter 1, by eugenic concerns and anxieties about the decline in the quality of the population revealed during the South African War and build up to World War I. Barclay Baron commenced the preface to his study of city boys and girls in 1911 with a quote from the 1904 Report of the Interdepartmental Committee on Physical Deterioration: 'We are impressed with the conviction that the period of adolescence is responsible for much waste of human material, and the entrance upon maturity of permanently damaged and ineffective persons of both sexes.' Concern about this 'wastage' would form the basis of much club work.[60] Baron referred to the poor health of many schoolchildren, with their defective vision, bad teeth, and incipient tuberculosis and heart disease.[61] In the factories which so many youth entered matters got worse, as they were exposed to long hours of heavy work and conditions of bad sanitation and ventilation, noxious fumes and ills such as lead poisoning. The 'monotony' of factory work also produced 'mental and moral disabilities', ill effects which 'the boy or girl, who is already undergoing the physical strain of "growing up," cannot be subjected to without suffering'. Quoting from one Factory Inspector's report, Baron described how in the textile industries girls exhibited ' "shortness of stature . . . miserable development . . . sallow cheeks and carious teeth". Even their hair . . . at an age when other girls have a luxurious growth, dwindles to a mere "rats tail".' The home for some young people was their 'best safeguard', but in others 'most unfavourable to the health of body and mind'.[62]

The Girl Guides, established in 1910, were renowned for their promotion of organised sports and games, intended not just to develop pluck, verve and general fitness, but also to address the poor health of Britain's children and anxieties about the deterioration of the race, evidenced by the 'preventable deformities' of tooth decay, adenoids, knock-knees, curvature of the spine and pigeon-breasts, all regarded by the organisation as endemic, particularly amongst town dwellers.[63] The Guides' *Handbook* for 1912 gave advice on how to build endurance, on exercise to improve the action of the heart, lungs, digestive system and each part of the body, and games to develop strength, including skipping, fencing, swimming, rowing, athletics, flag work and Morris dancing.[64] The Guides also strongly promoted health maintenance through 'cleanliness, temperance, and first aid'.[65] Lucy Freeman, turn-of-the-century pioneer of working girls' clubs and Catholic guiding in the Brighton area, argued that drill was 'an invaluable remedy for the narrow chests and bad carriage we see in so many girls', and also recommended dancing for girls once they had 'become civilized', as well as games, swimming, tennis, and walks and excursions.[66] Baron concluded that 'gymnastic exercises or musical drill are among the best resources of a girls' club', while 'in skill and, allowing for a smaller natural muscular development, in pluck and endurance, they are often not a whit behind boys in those sports which have been opened to them'.[67]

As with schoolgirls, team sport was recommended for working girls for its promotion of 'skill, dexterity, coolness and courage, with presence of mind. Cultivate and play all the sports you can in the open air, and they will make you a fitter type of perfect womanhood.'[68] Hockey, rounders and other organised team games were advocated in a 1913 number of *Girls' Club News* as valuable 'for the moral well-being of the working girl'. In response to an earlier article which expressed concern about the risk of hockey exacerbating the varicose veins which so many mill girls suffered from, the author argued that accidents were rare, but fatigue was more of a risk as 'some of the girls are of poor physique'.[69] The GFS encouraged hockey for its promotion of loyalty and unselfishness 'as a good hockey player always unites well with her own side, and never has a thought for herself'.[70] 'Outdoor games', a piece in the YWCA *Gazette* reported in 1920, 'are a valuable means of keeping girls in good health', as well as assisting in 'the development of self-control and self-forgetfulness'.[71] Beatrice Webb suggested that hockey was a suitable actively for girls 'giving concentrated exercise and the best kind of moral discipline, that involved in "Play the game!" – discipline of a kind which comes less readily to factory girls, each working on her own'.[72] In that sense, there was little to distinguish girls from the ethos advocated for factory boys or young women whose sporting exploits were conducted on the playing fields of the best girls' schools.

Many facilities, particularly well-stocked gymnasia, would be out of reach for working-class girls, and clubs noted time and again how difficult it was to find the space, equipment and teachers to facilitate sporting ventures. 'In the very hot weather we went to parks to play rounders, a favourite pastime' though it was 'not very easy to find an open space, free from rough, vulgar people, where these grown-up girls can disport themselves without attracting unpleasant attention'.[73] The YWCA and Girl Guides offered opportunities to girls with few openings to sport and leisure facilities. Agnes Baden-Powell's foreword to Dorothea Moore's novel *Terry the Girl-Guide* claimed that without organisations such as the Girl Guides girls of the lower classes would not improve physically or morally; the pursuits they offered appealed to all classes, including 'less favoured girls' who had no access to games.[74] As Mavis Kitching, who was born in Birmingham in 1916, recorded: 'Poor meant alongside running errands to earn a few pence, or fetching coke from the gasworks, being given a holiday by the Girl Guides... the guides paid for me, they all put together so that I could go with them, I liked it all the way through.'[75] Many activities, such as mass drill, became markedly more popular and accessible, with 3,000 girls taking part in a display at the new Wembley Stadium in 1923, 'a magnificent sight to those who have the future of the children of the Empire at heart'. 'If in an ordinary club match or competition self must be forgotten in team spirit, every ungenerous thought suppressed, and every ounce of *esprit de corps*, co-operation, carefulness, accuracy, self-control be brought forth, how much more so when we are drilling or playing for the whole Y.W.C.A. of our territory?'[76] (see Figure 5.2).

In addition to facilitating Empire goals, clubs offered girls shelter and meaningful recreation off the street corners that they were supposedly so attached to. They engaged too with the psychological impact of work and the detrimental effect that entering the workforce could have on wellbeing. Girls were depicted by the YWCA and other organisations advocating club activity, as facing numerous challenges, including loneliness and a lack of funds.

A girl may have left school at 15, full of health and vigour. Can she keep and develop that fitness when she begins long hours of sedentary work in the rush of town life? Nearly all Y.W.C.A. Clubs have on their programmes classes in physical drill, in gymnastics, and dancing, and the effect on health is often marked and rapid.[77]

A Convalescent and Holiday Fund was set up in 1884 and by 1887 had sent 950 city girls to cottages and lodgings in the countryside or by the sea for short breaks, as well as arranging day or Saturday afternoon walks and outings. By 1909 the numbers taking advantage of these holidays

Drill Competition at the Guildhall, London

Figure 5.2   Drill competition at the Guildhall, London, 1924.
Source: From *YWCA. A Review 1924*, p. 2 (Modern Records Centre, University of Warwick, MSS.243/2/1/9: YWCA/Platform 51).

had increased to 980 and the Fund also organised 34 outings in that year.[78] The Manchester Girls' Institute in Ancoats set up a Holiday Home in Southport, where about 400 girls were sent each summer, and additionally organised events and picnics.[79] The 1904 Report on Physical Deterioration encouraged a keen interest in activities that were both physical and outdoor, and youth work aimed increasingly to compensate for the stress of the city by taking boys and girls to the countryside at every opportunity.[80] Maude Stanley argued that girls were less likely than boys to venture to the countryside on their own initiative and had a greater need for country holidays: 'I cannot certainly say enough of the physical benefits of these country visits; but how much more there is to say on the moral aspects of these holidays!'[81] The National Organisation of Girls Clubs (NOGC) also coordinated holidays for working girls, which allowed girls from London and other large towns to

> obtain for themselves the fresh sea air, good food and early hours so necessary to young workers when health begins to fail. It is

considered by experts that work is in many cases carried on by the girls until, as they say themselves, they 'drop' and treatment often comes too late for thorough restoration of health.[82]

The YWCA, along with other organisations of girls' clubs, offered not only tips and guidance on health but also projected a rounded and, in many ways, 'modern' vision of the components of good health. This engaged with the psychological as well as the physical in taking the factory girl as a subject of reform in order to create a modern and healthy working girl. Claire Langhamer has argued that leisure cannot be separated from work in conceptualising women's experiences, and clubs recognised in undertaking reform work that working girls had strong feelings of entitlement to friendships and amusements.[83] The presence of girls in cities and towns, on their way to the factory or workshop – as well as their attachment to unmanaged leisure pursuits – also made them highly visible, particularly as their numbers increased. It was also recognised that such girls were likely to be economically independent and to manage their own lives without husbands or children for many years, and forming a bridge between the order of school and maturity of adulthood was crucial to club activity. After spending several winters in a remote part of London, Albinia Hobart-Hampden became convinced that club work was of 'supreme importance' for the future of the nation, particularly when the abrupt termination of public education left girls 'without any help or guidance at the most critical moment of her life'.[84] The Marple Girls' Institute in Lancashire, established in 1904 to pay particular attention to healthy recreation and instruction, worked closely with girls from the age of 13 who were employed in local cotton mills, suggesting that it offered the best practical solution on

> How to prevent the waste of time and moral deterioration which occurs between girls leaving school at fourteen and the time when the girl or boy either realizes that it does not pay to be ignorant and joins in an evening class, or takes his or her place amongst the unemployed.[85]

Deborah Valenze has outlined the construction of the factory girl in the mid-nineteenth century and the Victorian fascination with their moral condition and physical status. They were described as pallid and weak, 'poor decrepit creatures' at odds with middle-class ideals of prosperous, robust and blooming womanhood, and associated with indecency, coarse language, and wasting money on trifles, 'condemned for her immoral behaviour and pitied for her arduous workday'.[86] Such stereotypes persisted well into the early twentieth century, despite slowly improving conditions of work and living standards, and were even reinforced in some ways by club propaganda. For many commentators the term 'factory girl' meant 'a rough, rude, untidy, uncouth individual', though others disputed this, arguing that in some ways they expressed 'the chief charms and virtues of a good, refined womanhood'.[87] Maude Stanley, describing the activities of clubs for working girls, referred at some length to the difficulties of managing the girls' volatile and unruly behaviour, and the opportunities that sport offered to counteract these tendencies: 'At first they were hardly civilised, and had bad habits and conversation over which a veil must be drawn, but now they are well-behaved in the club, can be taken out on visits to other clubs, and are eager to take part in the next musical drill competition.'[88]

The dichotomy represented by the factory girl – as a victim and at the same time hopeless, silly and lax – continued to influence debates on their health, as the poor environmental conditions in which they found themselves were exacerbated by their behaviour. In 1910 Dr Elizabeth Sloan Chesser described the very poor working conditions for girls employed in factories without proper ventilation and cloakrooms which left outdoor clothing damp and dusty at the end of the day, while heat and moisture led to rheumatism, phthisis and other forms of tubercular disease. However, young women were also 'careless' about their clothes and food, going straight out of doors from the moisture and heat of the factory, and neglecting their diet. Many girls were anaemic and suffered from dyspepsia, but this was 'the natural result of strong tea, the bread and butter and tinned food diet they subsist on'.[89] Life in the factory, Chesser continued, meant not only physical degeneration but also moral deterioration: 'The monotony and dreariness of factory life accounts largely for the drinking and gambling which prove such temptations even to the younger girls. Indeed, the married women are said to exercise a bad moral influence on the girls in many ways.'[90] Chesser sought a remedy for this in girls' clubs, though not the 'party clubs', which involved saving up for 'treats' of cakes and sweets, or rum, tea or gin, or a mixture of all of these. She praised clubs, such as the GFS, for providing cheap dinners and healthy amusement for mill girls, places to meet in the evening, harmless recreations such as drill and games, and for counteracting the temptations of factory life.[91]

The *Girls' Club News*, published after 1909 by the NOGC – which was set up to further the spiritual, educational, physical and industrial welfare of working girls and to function as an affiliating body for girls' organisations – provides insight into a rather different set of attitudes towards working girls. As well as dispensing advice on various issues relating to health and wellbeing, its journal actively sought to encourage girls to engage with industrial matters and the reform and improvement of their workplaces and conditions, and to be involved in the management of their clubs. The journal provided a great deal of information on the organisation of conferences, training courses, lectures and competitions, on campaigns to open restrooms for girls and women at public events, on industrial news and legislation, and conference reports, and aimed to provide its readers with a sense of the history of girls' contributions to the passing of factory legislation. It explained the ways in which female Factory Inspectors had taken up the complaints of girls in the workplace, which had resulted in the banning of overtime for young women, and legislation to prohibit the carrying of heavy weights and to improve ventilation, heating and restrooms in factories.[92] Similarly, the YWCA's Industrial Law Committee urged girls to make individual or collective complaints against employers in breach of the law, and girls did indeed make requests for the provision of protective clothing, cloakrooms or seating, for the expansion of canteen facilities and improvements to ventilation. The Committee also assisted with cases where girls were owed wages, compelled to work overtime, or put in moral danger by their employers, and even with one case of wrongful confinement in a lunatic asylum.[93]

In addition, the *Girls' Club News* publicised information on club activities, many of which engaged in health promotion. Chelsea Girls Club offered its 100 members drill and swimming, and was associated with the Guild of Health, which offered health lectures and coordinated a Country Holiday Fund. St John-at-Hackney Club, set up as a counter-attraction to the streets and to get girls involved in church activities, opened five nights a week by 1913, and offered drill, skipping, dancing and socials in a hired hall. Honor Club, in addition to normal club activities, had a lady doctor visit the girls once a week to dispense medicine and advice.[94] The NOGC urged the setting up of restaurants for working girls in response to the worries of club leaders and girls serving on their committees about the inadequate food eaten by working girls. Turning the usual argument around, they suggested that it was improper food that led to a 'don't care' frame of mind, and consequently to indifference in striving for good wages and satisfactory conditions of work. They also pointed out that the girls disliked the lack of refinement in eating houses frequented by working men, 'and as the only alternative they go to a tea shop and order buns or cake or tea, food which does not give them the energy they need for their daily work'.[95]

Despite the generally positive attitude to girls' potential to actively improve their own environment and health reflected in the *Girls' Club News*, more critical voices were heard, based on the notion of the factory girl as a poor object, ill-equipped to take her place in society as a future mother. Dr Brown, Medical Officer of Health for Bermondsey, described how saddened he was that girls of his district would not help him in his attempt to fight disease and death, and presented a grim picture of a factory girl who 'is not wise enough to join a club and learn to be a happy housekeeper'. Dr Brown pointed out that girls were taught a certain amount of domestic economy before leaving public elementary school aged 14, but generally preferred factory work which offered better wages and more freedom than domestic service:

Once the girls are in the factory they promptly forget all the domestic teaching that they have learnt in school, unless they belong to some Club and attend its classes. They have bread and tea for breakfast, scraps of fried fish for dinner, and odds and ends for tea and supper . . . As they cannot spend their evenings in the overcrowded home they go to music halls or cinema shows.

Aged between 17 and 21, such girls were declared likely to enter an 'improvident marriage'. They had no idea, Brown added, how to lay out money, how to cook or keep the house. 'Scanty and unappetizing meals led to discontented husbands', and matters got only worse when the family began to arrive; 'if the unfortunate infant survives it does so by its own vitality'. More often it dies, contributing to the high death rates of Bermondsey babies.[96] Another article in the *Girls' Club News* referred to the importance of visits to factories in the dinner hour, which could involve lectures, outings and hymn singing. It added that 'Girls of this class have no initiative; often living within a penny ride of the park or country lanes they will never dream of spending a Saturday half-holiday anywhere than in their own familiar sordid street.'[97]

Club activities – and club literature – was intended to help equip girls for wifehood and motherhood, an objective advocated by club workers such as Lily Montagu and Maude Stanley and reflected in practical terms in published materials and courses on baby care and domestic skills. Some organisations, such as the GFS, placed a great deal of emphasis on domestic duty, mothercraft, cookery and sick nursing, as well as the practices of purity and self-control which would minimise the risks of adolescence. Yet, on the whole, girls' clubs appear to have been more immediately concerned with the needs of the current generation of adolescents as much as – if not more than – the training of future mothers of the Empire.[98] Notably, however, the message changed in the war years, when the YWCA and other clubs strongly urged the idea of service and stepped up their emphasis on acquiring and practising domestic and nursing skills and showing responsibility to others. The Girl Guides founded shortly before the outbreak of the First World War, when anxiety about degeneration peaked and as race 'elided with national identity', stressed the needs of Empire and the duties of girls to serve the common good.[99] Its mission was to urge self-education in character, handicraft, service for others, and health and physical development, and it was stressed that 'the girls of to-day are the home-makers of to-morrow, and also the mothers of the future'.[100] Girls 'can do a great deal for the country by learning the rules of health, by practising them personally, and by applying them to the care of children in their own homes, and teaching them to others'.[101] Yet, while guiding aimed to produce both good future mothers as well as physically robust girls, in so doing it also expanded the range of pursuits available to many young women.[102] Whatever its Imperial objectives, guiding offered access to a range of sports, new skills and outdoor opportunities to girls from a broad range of social backgrounds, including camping, orienteering and trekking.

Promoting a religious outlook and social purity was embedded in all club activity. However, many clubs, including the YWCA, did not insist on their members belonging to any particular denomination or on practising religion at all, though they certainly preferred it if they did. Others, such as the GFS, appointed only members of the Church of England as club leaders but encouraged girls of all denominations to become members. Club work also embraced the notion of the unity of moral, spiritual and physical welfare, suggesting that good health required commitment to all three. Flora Lucy Freeman advocated drill and gymnastics as aids to self-control as well as physical health. Her organisation also emphasised moral, spiritual and physical welfare and social purity.[103] 'Health' for the YWCA referred to the 'spiritual, mental and physical' and club work promised 'healthy' amusements, companionship, cheerful surroundings and a religious atmosphere, and encouraged 'pure recreation', complete relaxation as a counter to tedious work – 'a day of standing at the machine, sitting at the desk, or performing monotonous household duties'.[104]

The health challenges of the First World War

During the First World War large numbers of young women entered the workforce under emergency conditions and by April 1916 the female labour force had swelled by 600,000.[105] According to the Board of Trade, the number of women in non-professional jobs increased by 1,590,000 between July 1914 and July 1918, and some 891,000 of these jobs were in 'industrial occupations'.[106] The war exacerbated the poor conditions of employment for young women, particularly hours of work in munitions factories and exposed them to industrial diseases, though as Braybon and Summerfield have pointed out, women in traditional trades had worked for many years under considerable strain and in appalling environments, which received little attention.[107] Nonetheless, conditions in munitions factories were very poor and the hours of work long, in Dorothy Poole's case in an aeroengine works from 7 am to 5.30 pm, with two hours overtime; the factory was overcrowded, the food awful and the factory warmed only by buckets of coke.[108] Most visible and emblematic of the exposure of girls to disease was TNT poisoning, which resulted in jaundice and the distinctive yellow colouring of 'canary girls'; 'the girls turn yellow, and then many of them get horrible rashes, and their faces swell up so that they are for a day or two quite blind, and most repulsive objects. Nevertheless, when they are cured, they go back, and run the risk of getting ill again.'[109] Many of the women interviewed by Braybon and Summerfield were in their mid-teens when they worked in munitions factories and admitted that they did not appreciate the risks they were running when working with explosives.[110] The war also fanned anxieties about the sexual behaviour of young women excited by the atmosphere of war and often living away from home and led to concern about controlling venereal disease.[111] It served to magnify the image of young vulnerable women, hard-pressed and overtaxed at work, and living in dreadful conditions, and drove efforts to improve their welfare, health and fitness (see Figure 5.3).

In 1915 the Ministry of Munitions appointed the Health of Munition Workers Committee to consider questions of industrial fatigue, the efficiency of workers, hours of labour and the health of labourers in munitions factories. Welfare inspectors were appointed by its welfare sections to work in government-controlled factories and many of these positions were filled by women. Over 1,000 female industrial welfare inspectors were appointed, usually middle-class women who were largely supervising working-class women and girls.[112] Conditions started to improve after 1916, as proper toilets were installed (largely

At work on Munition Boxes.

Figure 5.3 At work on munitions boxes.
Source: From *YWCA. A Review 1916*, p. 16 (Modern Records Centre, University of Warwick, MSS.243/2/1/2: YWCA/Platform 51).

due to fears that poor hygiene would increase the incidence of venereal diseases), seating became more common, factories were built with restrooms, canteens and better ventilation, and medical facilities were introduced.[113]

As the government began to build shell factories and engineering shops at a feverish rate, the YWCA was called upon to assist and to address the 'unspeakable' conditions which resulted from the rapid creation of new work sites which were often in the middle of nowhere, and where the girls, who were fragile and 'out of harmony' with their conditions of labour and at risk from explosions and other occupational hazards, were compelled to forage in local villages for supplies and to eat outdoors, even in winter.[114] The YWCA Munition Workers' Welfare Committee was formed in June 1915 and by the time it was dissolved in February 1919 it had opened 101 centres. As requests poured in – to manage a colony of 700 girls, to start a canteen for night shift workers at 24 hours notice in a lonely part of Surrey, and to respond to a policeman imploring the YWCA to open a canteen to the thousands of girls who had 'invaded' his quiet village – the YWCA demonstrated impressive adaptability and speed of response, often working with other women's organisations, in setting up hostels and canteen facilities, rest and club rooms.[115] The London Division of the YWCA alone set up 21 hostels, three permanent canteens feeding thousands of girls each day, 21 clubs, several in slum districts, and additionally stepped up the organisation of drill teams and summer camps.[116] The association recognised that war work imposed enormous stress on young women: 'Long hours, overtime, meals in a crowded restaurant or shop, and journeys home in a still more crowded bus or tube, all tell on the nerves.'[117]

> On no class of the community has the physical strain of the war told more heavily than on the female munitions workers, mere children some of them, of fourteen or fifteen. Their working day is sometimes very long, sometimes they work in eight-hour shifts, which necessitates working one week out of three on night work.[118]

An appeal for support to set up hostels, clubs and restrooms stressed that, while the girls were 'working bravely', many were 'breaking down under the strain, always fighting through long hours of heavy work'.[119]

Canteens offering nutritious food at low prices were designed to improve health and comfort, as well as to fuel energy levels required for efficient work, and formed a cornerstone of the YWCA's activity. As Vicky Long has pointed out, canteens were viewed as a valuable mechanism to improve productivity, and were designed to serve nutritious food rapidly to large numbers of workers; they were thus often austere and utilitarian.[120] Yet the YWCA also saw their provision as providing the opportunity to instil domestic values and good eating habits in the girls who used them. It was reported how the good wholesome food was appreciated by the girls, as well as the neat tables, warm and cosy atmosphere and cleanly service (see Figure 5.4). Improving the environment where food was consumed and encouraging girls to eat a healthy diet, also aimed at improving their eating habits.[121] The girls apparently shared in this enthusiasm for canteen provision. In 1917, commenting upon a YWCA hut set up in northern France, Elsie Cooper commented that it was 'the nearest thing to home. We even had cups and saucers, not exactly china, but very superior to our usual tin mugs.'[122] Monica Cosens, a Southampton tram conductor, was also enthusiastic about YWCA canteen provision:

> At the far end are two counters, one piled with buns, oranges, sweets, lemonade; the other given up to urns of boiling water, mugs of tea, glasses of milk, whilst above it swings a large blackboard, on which is written the day's list of hot dishes prepared in the kitchen close at

hand, and the announcement that 'workers' own food will be cooked at the charge of one penny.[123]

Alongside canteens and centres for girls employed in factories building aeroplanes and making shells, hand grenades and ropes, clubs were set up to encourage outdoor activities such as tennis, swimming and cycling as 'a welcome and healthy change for the girls, many of whom have been working...at arduous and monotonous tasks'. Lectures, debates and study circles were also initiated.[124] The WYCA was keen that this work continued beyond the war years, which had given 'a great impetus' to physical culture and had also seen the acquisition of land for sports facilities and allotments.[125] Holidays were offered to increasing numbers of girls, and during 1918–19, 10,679 girls enjoyed a holiday in a camp or hostel run by the YWCA.[126]

During the war clubs focused increasingly on issues of citizenship, which centred largely on developing a positive approach to work and the work environment and encouraging girls to understand the contributions they made to production and the war effort. The *Girls' Club News*

Y.W.C.A. Munition Workers' Canteen

Figure 5.4   YWCA munition workers' canteen.
Source: From *YWCA. A Review 1914–1920*, p. 5 (Modern Records Centre, University of Warwick, MSS.243/2/1/5: YWCA/Platform 51).

referred to how girls had responded to the needs of war by switching to employment entailing long hours, difficult conditions and dreariness, but also reminded girls to be responsible and to avoid 'disgracing womanhood', caught up as they were in the excitement as much as the drudgery of war. The role of clubs in this context became more important, enabling girls to lead as normal a life as possible.[127] Lectures offered by clubs reflected on the findings of the various war committees on women's employment and munitions work, and provided tips on how to keep well, take care of teeth, skin, hair, and appearance, on healthy attire, how to avoid indigestion and consumption, as well as discussing wifehood and motherhood, and girls' responsibility as citizens. Stepped up engagement with girls' future role as mothers was an unsurprising feature of the war, given the enormous loss of life coupled with continuing high levels of infant mortality. Beatrice Webb commented in the introduction to her manual for welfare supervisors published in 1917 how the health of working girls had engaged public attention, but during war 'a higher value was put on the working woman':

> Her importance as the potential mother whose function it may be to make good the fearful loss of human life which has taken place during the past three years, and the part she has to play in producing munitions, equipment and food, and exports, have become recognized, and have brought home the fact that consequently every effort must be made to keep her strong and in good health, and fit for the work she has to perform.[128]

Webb's volume trod a fine balance between urging provisions that addressed the health and welfare of the girls and minimised the risks of overstrain yet also ensured maximum effort and productivity. She captured the need for both physical activity and mental wellbeing, recommending free time to be devoted to 'a vigorous game or dance in the open air ... Ten minutes spent in full activity in the factory garden or yard sends the girls back refreshed and quickened up to an extent which shows itself in the increased output of the next hour.' Good canteen arrangements were vital 'to save every second of time', but also to avoid pushing, annoyance and mental unrest. At the same time long hours of strenuous work could cause special difficulties as regards the reproductive organs: 'Cases in unmarried girls of prolapse of the uterus ... have been increasingly common since the war began ... Heavy lifting, sudden strains in pushing or pulling, especially in the case of girls who wear tight corsets ... account for most of these cases.'[129] Here, Webb managed to weave in the notion that not only environment and work but also girls' behaviour and habits caused their health problems. Constipation was highlighted as a commonly reoccurring condition and Webb castigated girls for spending money on proprietary medicines to treat this: 'If the money so spent went on green vegetables and fruit, or were saved up for a bicycle, while the girls attacked constipation by means of a regular habit, suitable foods, drinks, and exercise, it would be vastly more to the point.'[130]

It was acknowledged that workplace conditions were poor, girls were not in an optimal state to conduct industrial processes, and that industrial diseases and illness triggered by long hours of work in poor conditions were rife, yet it was also argued that the war, not least improved wages, could improve health. Dorothy Collier's 1918 report on *The Girl in Industry*, which cited evidence collected from girls' club secretaries, described anaemia as rife amongst exhausted girls, but also pointed out that some welfare workers believed that anaemia had reduced during the war as a result of better diet and a more regular lifestyle. A similar ambivalence was expressed about nervous disorders. Some doctors, Collier reported, related these to boredom and restricted movement which could make 'a whole room of girls... nervous and hysterical', while others claimed that the social influences at work, companionship and increased interest in life, served as antidotes to nervous afflictions. In terms of girls' physical health, painful or excessive menstruation, Collier concluded, was rare, though some girls found continued standing tiring during menstruation, while particular movements could also result in trouble with the womb and bowels. Like Webb, she referred to the prevalence of gastric disorders which she concluded were unrelated on the whole to industrial work; rather, tea and buns were again implicated, as well irregular and hurried eating. Collier also recommended active movement to offset the fatigue which seemed to affect girls mentally rather than physically.[131]

In 1917 munitions nurse, Sarah MacDonald, published *Simple Health Talks with Women War Workers*, intended for the munitions worker and those interested in her welfare. MacDonald urged open air activities as a counter to monotonous work and as an aid to support appetite, digestion and nerves. She offered extensive advice on hygienic practices to girls: 'So please, girls, don't spare the soap and water, both for your own sake and for the sake of the public at large.' With regards to hair, 'Some grown-up women are worse than children in this respect, which is rather humiliating, you know, and the things do breed, don't they?' 'Next in line are the teeth, and the number of girls with bad teeth is appalling.'

Fatigue and anaemia, with its associated headaches and nervousness, was attributed to badly ventilated rooms, lack of sunshine and fresh air, but also strong tea and improper diet. 'Meals should be taken at regular times', MacDonald instructed, 'and not in snatches, and sufficient time spent over the meals for thorough mastication in readiness for digestion'.[132] Such advice – directed at disciplining girls' bodies and behaviour – was not always well taken:

> Factory girls are just as sensitive to home truths as other women. They object to comments on their dress and their domestic arrangements from the welfare workers as much as she would object to similar comments from them. One lady made her debut in the factory where very rough girls worked by saying to them 'you want a club, you come from such overcrowded, dirty homes,' and then she was astonished when they threw their lunch at her'.[133]

Club activities aimed to reduce the impact of the monotony of work and the potential for moral lapses. The YWCA's affiliated branches of Girl Guides, claimed, for example, to do a great deal 'to bring the "flapper" into touch with the Association', and 'to attract the younger girl, whose life has suddenly become so unrestrained and so full of temptation'.[134] The conditions of war, night work, travelling unsupervised, living in hutments where they were exposed to corruptible influences and individuals, increased spending power, and girls' intrinsic gullibility, all exposed them, according to the YWCA, to moral temptation. Lectures delivered by the YWCA's Moral Education Sub-Committee sought to resist this, emphasising the dangers of venereal disease. 'The need for greater knowledge of this subject becomes daily more apparent, and people are realising at last that many tragedies might have been prevented if young men and women had been taught sooner about the laws of God which govern their physical nature.'[135] Reception hostels were set up by the YWCA for girls in 'moral difficulties', including a maternity home for single girls. Seventy-six girls passed through the home during the war, of whom 48 found situations in service either with the child or where the child was placed with a foster mother, 15 returned to their parents and five married.[136]

Sarah MacDonald elaborated on venereal diseases in her health lectures, addressing 'the thoughtless, frivolous, and unprotected' and pointing out that 'when you lose sons and daughters in sickness and in war…ah! Then you know that procreation…is not a joke.' She went on to explain the reproductive process and to instruct girls in menstrual hygiene, and also warned them against overstrain, urging them to study their abdominal muscles and to learn the knack of lifting heavy weights.[137] Eugenicist physician, Dr Mary Scharlieb, offered lectures to the YWCA on a variety of subjects, including venereal disease. She also published several booklets under the auspices of the Council of Public Morals for Great and Greater Britain and the National Council for Combating Venereal Diseases, expressing concern about the temptations of intemperance, impurity and want of self-restraint in times of war, and urging women and girls to actively contribute to the war effort and national regeneration. Girls were urged to understand the value of their bodies and the sad consequences of their misuse, to avoid drink which took away self control, and she warned of the risks faced by promiscuous girls – abandonment by their parents and by men who would leave them to face the perils of childbirth alone and fail to provide for themselves and their babies. Venereal disease, meanwhile, she explained, could cause miscarriage or stillbirth, or the birth of babies suffering sores and rashes, who were liable to die from wasting or become deaf, blind or paralysed and insane during their youth. Girls as young as 13 to 14 had become infected and 'the heaviest share of trouble falls on the girl'.[138] Scharlieb was keen to urge public education and overcome the 'policy of silence', deploring the fact that 'a certain proportion of children, especially of girls, has been shielded from knowledge of sexual physiology and hygiene. They have been allowed to marry and to become mothers without knowing anything definite and practical about the facts of life. The results have been disastrous.'[139] Advocating improved knowledge of sexual physiology and hygiene, Scharlieb stressed the need to raise standards of physical cleanliness, moral purity and self-control, and, particularly in times of war when temptation was rife, urged self-restraint, chastity and avoidance of drink.[140]

Conclusion

None of these challenges to girls' wellbeing were new, but they were heightened during wartime, as were other risks to health. Working conditions had slowly improved from the late nineteenth century onwards as employers inaugurated welfare facilities and fell in line with legislation managing hours and conditions, and took measures to improve 'the cleanliness, discipline and moral tone of the workplace'.[141] Interest in the health of working girls magnified during the First World War, as the numbers employed surged, particularly in munitions factories, which exposed them to the attention of welfare supervisors, who provided information on the particular diseases and disorders most likely to afflict working-class girls and also suggested ways in which they might improve their health and wellbeing. The bodies of working-class girls were scrutinised as never before in the workplace itself rather than in club rooms, as were their behaviour and attitudes towards work. Responses to health concerns by government, employers and voluntary agencies – the provision of canteens, improved hygienic facilities and working conditions, and efforts to support exercise and mindful recreation – might well have acted as a 'foil' to mask the fact that work in a factory environment was deeply damaging in the short and long term to a girl's health and future wellbeing, and it also provided the opportunity to impose rules dictated by middle-class values on working-class girls. While it was recognised that girls were pushed hard, particularly in times of war, they were described as having an enduring capacity for heavy and long hours of labour, something confirmed on the whole by the reports of the Industrial Fatigue Board and Medical Women's Federation towards the end of the war. It was also generally concluded that war work would not damage girls' reproductive wellbeing, though the moral impact of war work and its associated freedoms certainly could.

The First World War impacted on the tone and content of medical advice literature, heightening concern about girls' contributions to the future of the nation, as well as fears about falling standards in times of war. It was suggested that 'as the glamour of the patriotic appeal wears off.... she finds that war work is just the same old drudgery as repetitive work has always been'; in combination with increased wages, being herded with other girls to munitions camps, and having no connection with home life, made girls thoughts stray to 'a wilder, freer existence'.[142] In that sense, girls' reckless or inappropriate behaviour was described as overriding the impact of poor conditions or the work itself. Yet, increasingly, the psychological effect of work and monotonous tasks, long hours coupled with domestic drudgery and the absence of pleasure and pleasing environments for working girls was also conceived of as being destructive of health.

Health played a vital role in constructing a model of citizenship, particularly as good habits of health were reckoned to be formed by steady, daily effort. 'Cleanliness, plenty of exercise, erect posture, and a sufficient amount of sleep are all points which should be included and constantly emphasized in any plan for broad Club activity'.[143] The necessity of mental stimulation and fulfilment was also emphasised though the basis of this had been laid down already in club work. In 1918 in an essay on 'The Young Factory Girl', Emily Matthias explained how the 'immorality' of the factory girl could be put down to 'sheer reaction. She is drugged with monotony and long hours of physical labour, and feels the need for a strong sharp stimulus;' this required the creation of rational methods of letting off steam, as well as rest and exercise.[144] Helen Ferris pointed out that there were girls employed in factories 'whose powers of relaxation and sociability have often been dulled by the mechanical work in which they engage; the hours of fun and games at the Club may supply just what these girls most need'.[145] Education was urged as a stimulus to self-improvement and better citizenship, and was seen as intrinsic to the mission to improve health. Matthias also pointed out that girls were naturally houseproud and craved a pleasant, bright environment at work: 'such little things as growing plants and window boxes make a vast different to aesthetic enjoyment'.

> The young factory girl craves food of all descriptions – mental, phys-ical, aesthetic and moral food…The best the factory employer can do is to give her essentially good conditions, opportunity to assume responsibility according to her powers, technical education so that she may do the more intelligent jobs, sufficient skilled and sym-pathetic guidance; and then to leave her alone to find her own directions of development.[146]

In considering the appearance of their hostels, the YWCA explained that 'if environment counts for as much as modern psychologists insist, they must be as artistic and cheerful as possible'.[147] The organisations' colleges, meanwhile, taught working girls a range of skills, including household efficiency and how to reinforce the natural mother instinct, but also languages, drawing and nice ways of speaking. Their object was to help girls from working-class homes 'transferred straight from school to work that is probably monotonous, who suffer from mental cramp that prevents their growth'.[148] In the 1930s this work was extended in industrial study camps which encouraged girls to understand the work that they were engaged in, made them aware of their contributions, and thus claimed to raise morale and overall wellbeing. Yet the continuing challenge of adolescent girls was still being referred to in 1932 in a hand-book for girls' club workers. Girls were depicted as harder to understand than boys, more elusive and sensitive, 'strange physical changes' made them seem remote and moody, they could be swanky or tomboys and insubordinate girls posed special problems. It was also pointed out, how-ever, that much more was demanded of girls who had home duties after a day at school or in the factory and were often simply exhausted.[149]

Club activity appears to have been genuinely concerned to improve the health of working girls and evolved broad definitions of health and creative remits in improving it. Organisations, including the YWCA, Girl Guides and GFS, lauded maternity and domestic skills, as well as the needs of nation and Empire, and this aspect of girls' lives certainly gained additional resonance during the war years, spurred on, as seen in previous chapters, by eugenic concerns and eugenic writers and publicists. The YWCA was keen to promote healthy motherhood, and believed factory work in the long term to be detrimental. Yet girls' clubs as well as welfare supervisors appeared to engage little with the notion that girls were bounded by their physiological or biological limitations, something referred to infrequently in club literature. The YWCA was also greatly concerned about the immediate health needs of adolescent girls as independent workers in terms of their physical, mental and spiritual wellbeing. Citizenship was not just about motherhood and contributions to the nation, but also about girls understanding their role in society, enjoying work and combining it with sensible and meaningful recreation as much as domestic skills, and developing a sense of worth, rather than being tied chiefly to anxieties about their role as future mothers of the Empire.

6
Conclusion: Future Mothers of the Empire or a 'Double Gain'?

The Victorian period was at one and the same time notable for its deep medical conservatism, marked by the publications of vocal male physicians on the biological limitations of young women, but also for undercurrents leading to a change in attitude towards girls' potential to improve their health and physique. Debates and advice on healthy girl-hood, invoking new visions and new practices of health, would come to fundamentally shape the lives and lifestyles of adolescent girls. During the late Victorian period a number of medical authors and, increasingly, new forms of writing about girls' health, including a vast array of prescriptive material, conveyed increasingly positive health messages. While medical literature still emphasised how risky and unpleasant puberty and pubescent girls could be, given the restraints of biology and female nature, prescriptive literature tended to deal more pragmatically with the challenges of female adolescence and foster creative ways of achieving a healthy physique and lifestyle, based on attention to hygiene, exercise and diet, a balance between mental and physical work, and a positive attitude.

> The best exercise is that which combines some special object to be attained with the exercising of the muscles. The mind is thus stimulated, and consequently muscles and nerves work harmoniously together. Thus useful labour and games having some other object than mere exercise, neither over-tax the nerves nor exhaust the will power.[1]

Anxieties about young women's limited funds of energy, though durable, were steadily replaced with the view that energy could be creatively marshalled or even increased by healthful activities. While puberty needed to be scrupulously managed, by the turn of the century it was less likely to be described as disabling.

Across the relatively brief period surveyed in this book – the 1870s to the 1920s – girls were transformed in terms of dress, appearance and physical stature, as well as dietary and hygienic practices. They were exposed to new ideas about health on paper as well as in practice in the spaces they increasing inhabited outside of their homes. Rather than resisting their changing roles, the authors of the new literature and those engaged in work with young women, appear in many cases to have actively sought to respond to them, and even to have forefronted them, redefining girls as fit, active and capable of making new forays into education, work, sporting and social activities. Indeed, these activities were presented as being health giving, providing opportunities for the development of bodily and mental wellbeing. One downside to this was that girls' health was also increasingly judged as being shaped by their behaviour and attitude to life. If they failed to manage their health or their mindset the consequences could be dire and, whatever their social class, girls to a large extent were declared increasingly responsible for their own wellbeing, even if they were guided by headmistresses, factory welfare supervisors or via advice literature. Another downside was that large numbers of working-class girls would have only very limited access to opportunities to improve their health.

A range of experts interested in the health and welfare of young women devoted a great deal of attention to this topic. Some of these experts needed to rely on persuasion to engage girls in their visions of improved health, as with the working girls who attended clubs or the Girl Guides, yet access to even limited facilities for enjoying new sports and healthy activities in environments that were far more pleasant than their home or workplaces must have been alluring to many girls. A small number of accounts by girls themselves indicate that many aspired to take up exercise and other practices of health, even if their resources were limited. Thus we have Bim Andrews' regret that she did not have the wherewithal to play tennis and her purchase of a bike as soon as she could afford it, and Mavis Kitching's pleasure at being funded to join in guiding activities even if at the same time it branded her as 'poor'.[2]

The production of novels and magazines for girls has been extensively surveyed by authors such as Mitchell and Smith, with Smith concluding that this period was marked predominantly by a concern with Empire and attention to the ways in which girls could contribute to the national good and act as buffers against race deterioration.[3] With its emphasis on exploring a variety of different publications on health, including prescriptive literature, the findings of this study nuance this argument. A broader set of objectives came into play in the promotion of good health amongst girls, with doctors, headmistresses, feminist eugenicists, club leaders, and a variety of other writers speaking to complicated societal and political concerns about the emergence of the new cultural category of girlhood. It can also be argued that novels of the period were more likely to promote the Empire's interests, keen as they were to explore themes such as adventure and heroism rather than more mundane health maintenance and improvement.

Healthful activities, including sport, were construed – depending on who was making the assessment and when – as potentially improving girls' prospects as mothers but also as potentially imperilling maternal function. The possibility of a 'double gain', associated with girls who were keen to train their bodies and minds, but who were also likely to become excellent housewives and mothers, gained purchase in the late nineteenth century. At the same time, many of those advising girls on health matters appeared to be predominantly interested in the more immediate gains of healthy girlhood. Yet, with birth rates in decline around the turn of the century, at least for affluent and well-educated members of society, and the threat of war and war itself, anxieties reignited about girls' future capacity as mothers. This dovetailed with the rise of eugenics and with heightened concerns about adolescence as a time of both potential and great risk; for girls this was often expressed in terms of their abandonment of traditional femininity and acquisition of masculine characteristics. Even many headmistresses, who had been so keen to improve girls' health and capacity for study, now regretted that sporting activity had gone too far, resulting in overathleticism and the demise of feminine attributes. In 1911, eugenicist physician Mary Scharlieb described gymnastics and games for girls as still useful in building health and character, but remarked that doctors and schoolmistresses observed that 'excessive devotion' to athletics and gymnastics had produced flat-chested and narrow-hipped girls, their suitability for motherhood dissipated.[4] Mrs Eric Pritchard reminded readers of the *Pall Mall Magazine* at the end of the South African War, as the infant welfare movement was building momentum, that physical culture was rendering women unfit for maternity. Women were developing bodies like men, especially from the waist downwards, she argued, though uncontrolled athleticism.[5]

Concerns about the number and quality of future generations also permeated advice literature around the turn of the century, as it focused increasingly on instructing girls in domestic subjects and baby care.

In her introduction to Annie Burns Smith's, *Talks with Girls upon Personal Hygiene*, published in 1912, Mary Sturge commented on the 'lamentably low' health of large numbers of the population, and the importance of 'training our girls in the laws of life and growth, as well as in home duties and hygiene'.[6] Smith's volume, intended primarily for elementary schoolgirls, included several chapters on baby care, notably infant feeding. While presenting a modern image in terms of its interest in promoting a scientific approach to personal, domestic and public hygiene, and methods of mothering, in many ways the book harked back to the idea of 'little mothers', except that these little mothers would not be taken in hand by their own mothers or older siblings, but by experts who trained them for the future task of producing healthy babies, reflecting the broader incursions of physicians in promoting scientific motherhood and intervening in a wider range of health matters.[7]

Thus the Stenhouses, in their *Health Reader for Girls*, published in 1918, suggested that infants who were raised by inept mothers were in far more danger than soldiers at the front.[8] Their approach to imparting information was, however, more subtle than the overall message that it was intended to convey, as they slipped in instructions on baby care between advice to girls on health, diet and hygiene. Thus a chapter on personal cleanliness moved seamlessly from the care of the hands, skin and hair to 'the baby and cleanliness', a chapter on clothing offered advice on choices of fabrics, the dangers of tight-lacing and bad footwear, the care of clothing, and the clothing of babies. Though ostensibly a health guide for girls, the authors declared that 'The main object of this little book is to show how, in a very real sense, we are always at war, and to make it clear to girls that in this great struggle against the enemy we rely chiefly upon them to win the victory.'[9] The enemy was poor health, particularly when it affected the youngest members of society. Girls' roles as future mothers and homemakers were underlined. They needed to know about the human body, and the wise use of food and beverages, and the dangers of dirt and foul air, as well as the special needs of babies and young children, as 'it will be their duty and pleasure, as wives and mothers, to select and prepare the food of their families, to see that houses are kept clean and airy, and to care for the children who will become the men and women of the future'.[10] This 'should make a girl very proud and fill her with eagerness to learn all she can of the laws of health...to lay a foundation of health and happiness for herself and future generations'.[11]

Duty to country and the national good was powerfully expressed in the opening remark of Mary Humphrey's volume, *Personal Hygiene for*

Girls, published in 1913: 'Only a girl! But what an important thing it is to be just that – only a girl! If it were possible to make every girl realise how really important she is, how serious and responsible a thing it is to be a girl, **the future welfare of the nation** would be assured.'[12] Ideals of citizenship strongly pervaded health literature in the years prior to and during the First World War, and many guides suggested that once they had passed beyond 'girlhood', healthy marriage, home management and motherhood awaited. The preface to Humphreys' book, written at the threshold of war, declared that on 'the full powers of body, mind, and spirit of the girl of to-day...the future well-being of the individual, the prosperity of the community, and the peace and security of the nation mainly depend'.[13] Good health was a patriotic and moral duty: 'Neglect of health is a particularly selfish and absolutely unpardonable offence...*Keep well* for your own sake, for the sake of the nation which looks to its youth for the regeneration of its general health.'[14] Preparedness for womanhood was a serious obligation: 'What is looked for now is strong, capable, efficient women to carry on the traditions of our race.'[15] Yet good health was also described as an end in itself, and healthy girls described as able to contribute to the national good in a variety of ways, through voluntary work, industrial labour, the professions, competitive sport and so on. Much of Humphrey's book of health, hygiene and beauty in fact made little direct reference in terms of its content to girls as future mothers, while the image selected for the frontispiece depicted a breezy modern girl in fashionable attire, ready, not to tackle baby care, but a game of golf (see Figure 6.1).

Seemingly unequivocal commitment to maternity was expressed by eugenicist physician Mary Scharlieb:

> Speaking specially with regard to girls, let us remember that the highest earthly ideal for a woman is that she should be a good wife and a good mother...she ought to be so educated, so guided, as to instinctively realise that wifehood and motherhood is the flower and perfection of her being.[16]

Fellow eugenicist Dr Elizabeth Sloan Chesser insisted in 1914 that girls should be provided with basic training in the management of children as a crucial aspect of the eugenic mission, 'encouraging healthy and worthy parenting and the better care of offspring'.[17] Girls and boys, she argued, should be trained to preserve their health and for good parenthood. Girls additionally 'should be given simple training in the management of children, because, whilst all girls will not become mothers,

"In flower of youth and beauty's pride."—*Dryden.*

Figure 6.1 Mary Humphreys, *Personal Hygiene for Girls* (London, New York, Toronto and Melbourne: Cassell, 1913), frontispiece (author's collection).

few women will go through life without having had something to do with the care of children'.[18] Her book served as a broad health compendium for girls, and can also be envisaged as part of a much bigger project to instil in girls the need for a scientific approach to hygienic feeding practices and baby care, which was part and parcel of the campaign to raise standards of mothering through mothercraft courses and advice literature.

Yet by this time girls had asserted themselves in a number of new spheres, backed by feminists, headmistresses, social commentators, and club leaders as well as medical professionals. Female eugenicists faced something of a dilemma when advocating pro-natalist agendas dependent on promoting motherhood in which young women played a crucial role. As Greta Jones has suggested, attention to motherhood might end up leading to restrictions on women's lives, as well as excessive state interference.[19] While not losing her focus on the importance of housecraft and training for motherhood, particularly given the rising age of marriage, Mary Scharlieb continued to advocate careers for women, which could involve university degrees or other higher examinations, training for the Civil Service or as teachers, in the arts or music, while Elizabeth Sloan Chesser supported women's engagement in 'real' employment and social and voluntary activity in addition to work in the home, with good health and vitality supporting such varied pursuits. Girls should aim to harness good health and mental health 'to cultivate industry, the love of work, the capacity for application'.[20] Dr Christine Murrell saw motherhood as part and parcel of the female life cycle, but also urged girls to be ambitious and not to regard marriage as the end of training or a career. She also made the point that menstruation was normally not disabling, which was underscored by the Medical Women's Federation from the 1920s onwards.[21]

By the turn of the twentieth century, girls had already taken up secondary education and entered the workforce in increasing numbers, and they were joining sporting and girls clubs, involving themselves in a wide range of exercises and other social and recreational pursuits. At the same time, it could also be argued that girls themselves became precious commodities, as families diminished in size and with the increased emphasis on adolescence as a time of potential as well as challenge. This coincided with an enhanced interest on the part of many households in improving the health of all their members. Viviana Zelizer's account of the 'profound transformation' of the economic and sentimental value of children between 1870 and 1930 has been applied particularly to the US, young children and child labour. Her thesis that children were rendered economically useless through their transformation from workers into scholars, and, at the same time, emotionally priceless to their parents and wider communities, has some purchase, as Carolyn Steedman has demonstrated, with regard to the establishment of compulsory schooling in Britain.[22] Zelizer's account also has relevance to the investment in the health of young women in Britain over a similar period, a time productive of a vast literature addressing this subject, and marked by the serious engagement of a wide range of authorities and agencies, for the great part private or voluntary, in promoting girls' wellbeing. This was in part to achieve economic objectives, in part to improve the prospects for mothers of the future, in part to fund schools or gymnasia or to sell copy, but also because – as with their brothers – girls were seen increasingly as having the right to good health, and this health was not just about physicality and not getting ill but mental and emotional wellbeing too. The potential for a 'double gain' also acquired considerable traction as a means of aligning the notion that girls had the ability to fulfil a variety of roles alongside their preparations for future motherhood – though it was also acknowledged that many would not marry, something which became ever more evident after the carnage of the First World War. Girlhood was conceived as a period of great opportunity for the individual and the nation, and it was recognised that many women would continue to make other contributions even if they were to become mothers. This 'double gain' perhaps was also crucial in laying down the 'double burden' of economic and domestic work that would mark many women's lives in modern Britain.

Notes

Introduction

1. Gordon Stables, *The Girl's Own Book of Health and Beauty* (London: Jarrold and Sons, 1891), preface.
2. Ibid., pp. 12–13.
3. Ibid., pp. 16–17.
4. See G.S. Woods, rev. Guy Arnold, 'Stables, William Gordon (1837–1910)', *Oxford Dictionary of National Biography* (hereafter *DNB*): [http://0-www. oxforddnb.com.pugwash.lib.warwick.ac.uk/view/article/36229, accessed 27 January 2012].
5. See, for example, Birgitte Søland, *Becoming Modern: Young Women and the Reconstruction of Womanhood in the 1920s* (Princeton, NJ and Oxford: Princeton University Press, 2000), Chapter 2; Ina Zweiniger-Bargielowska, *Managing the Body: Beauty, Health, and Fitness in Britain, 1880–1939* (Oxford: Oxford University Press, 2010), Chapter 6; David Fowler, *Youth Culture in Modern Britain, c.1920–c.1970* (Houndmills: Palgrave Macmillan, 2008), Chapter 3; Adrian Bingham, *Gender, Modernity, and the Popular Press in Inter-War Britain* (Oxford: Clarendon, 2004), Chapter 2.
6. Zweiniger-Bargielowska, *Managing the Body*, p. 109.
7. Sally Mitchell, *The New Girl: Girls' Culture in England 1880–1915* (New York: Columbia University Press, 1995), p. 1.
8. Mary Anne Broome, *Colonial Memories* (London: Smith, Elder, 1904), pp. 293, 295, 300. Cited ibid., p. 3.
9. Robert Roberts, *The Classic Slum: Salford Life in the First Quarter of the Century* (Penguin edn, 1973; first published Manchester: University of Manchester Press, 1971), p. 201.
10. Barbara Harrison, *Not Only the 'Dangerous Trades': Women's Work and Health in Britain, 1880–1914* (London: Taylor & Francis, 1996), p. 5.
11. Mitchell, *The New Girl*, pp. 1, 7.
12. David Fowler dates the emergence of a distinctive teenage culture to the interwar years in *The First Teenagers: The Lifestyle of Young Wage-Earners in Interwar Britain* (London: Woburn Press, 1995), p. 1.
13. Stables, *The Girl's Own Book of Health and Beauty*, p. 195.
14. Roberts, *The Classic Slum*, pp. 51, 53.
15. Elizabeth Roberts, *A Woman's Place: An Oral History of Working-Class Women 1890–1940* (Oxford: Basil Blackwell, 1984), p. 39.
16. Claire Langhamer, *Women's Leisure in England, 1920–1960* (Manchester: Manchester University Press, 2000).
17. Mitchell, *The New Girl*, p. 9.
18. E.J. Tilt, *On the Preservation of the Health of Women at the Critical Periods of Life* (London: John Churchill, 1851), pp. 36–7; George Black, *The Young Wife's Advice Book: A Guide for Mothers on Health and Self-Management*, 6th edn (London: Ward, Lock and Co., 1888), p. 5.

19. E.B. Duffey, *What Women Should Know. A Woman's Book about Women. Containing Practical Information for Wives and Mothers* (Philadelphia, PA: J.M. Stoddart [1873]), p. 28; Lionel Weatherly, *The Young Wife's Own Book: A Manual of Personal and Family Hygiene* (London: Griffith and Farran, 1882), p. 27.
20. Mary Hilton and Maria Nikolajeva, 'Introduction: Time of Turmoil', in *idem* (eds), *Contemporary Adolescent Literature and Culture* (Farnham, Surrey: Ashgate, 2012), pp. 1–16, on p. 2. G. Stanley Hall, *Adolescence: Its Psychology and the Relation to Physiology, Anthropology, Sociology, Sex, Crime, Religion and Education* (New York: D. Appleton, 1904) and *Youth: Its Education, Regimen, and Hygiene* (New York: D. Appleton, 1906). His ideas are explored more fully in Chapter 1.
21. Mary Scharlieb and F. Arthur Sibly, *Youth and Sex: Dangers and Safeguards for Boys and Girls* (London: T.C. & E.C. Jack, 1919), Part 1: 'Girls', by Mary Scharlieb, p. 7.
22. See Hilary Marland, 'Women, Health and Medicine', in Mark Jackson (ed.), *The Oxford Handbook of the History of Medicine* (Oxford: Oxford University Press, 2011), pp. 484–502.
23. Though for the US, see Joan Jacobs Brumberg, *The Body Project: An Intimate History of American Girls* (New York: Vintage Books, 1998) and Martha H. Verbrugge, *Active Bodies: A History of Women's Physical Education in Twentieth-Century America* (Oxford and New York: Oxford University Press, 2012), especially Chapters 2 and 3.
24. See, for instance, Joan Jacobs Brumberg, *Fasting Girls: The History of Anorexia Nervosa* (Cambridge, MA: Harvard University Press, 1988).
25. Patricia A. Vertinsky, *The Eternally Wounded Woman: Women, Doctors, and Exercise in the Late Nineteenth Century* (Urbana and Chicago, IL: University of Illinois Press, 1994; first published Manchester: Manchester University Press, 1989), quote on p. 39.
26. Lorna Duffin, 'The Conspicuous Consumptive: Woman as an Invalid', in Sara Delamont and Lorna Duffin (eds), *The Nineteenth-Century Woman: Her Cultural and Physical World* (London: Croom Helm, 1978), pp. 26–55; Maria H. Frawley, *Invalidism and Identity in Nineteenth-Century Britain* (Chicago and London: University of Chicago Press, 2004). See also Miriam Bailin, *The Sickroom in Victorian Fiction: The Art of Being Ill* (New York: Cambridge University Press, 1994).
27. Duffin, 'The Conspicuous Consumptive', p. 32.
28. Michelle J. Smith, *Empire in British Girls' Literature and Culture: Imperial Girls, 1880–1915* (Houndmills: Palgrave Macmillan, 2011).
29. See, for example, out of the vast literature on maternal and infant welfare, Jane Lewis, *The Politics of Motherhood: Child and Maternal Welfare in England 1900–1939* (London: Croom Helm, 1980); Deborah Dwork, *War Is Good for Babies and Other Young Children: A History of the Infant and Child Welfare Movement in England 1898–1918* (London and New York: Tavistock, 1987); Anna Davin, 'Imperialism and Motherhood', *History Workshop Journal*, 5 (1978), 9–66; Valerie Fildes, Lara Marks and Hilary Marland (eds), *Women and Children First: International Maternal and Infant Welfare 1870–1945* (London and New York: Routledge, 1992); Seth Koven and Sonya Michel (eds), *Mothers of a New World: Maternalist Politics and the Origins of Welfare States* (New York and

London: Routledge, 1993); Gisela Bock and Pat Thane (eds), *Maternity and Gender Policies: Women and the Rise of the European Welfare States 1880s–1950s* (London and New York: Routledge, 1991), and for the US, Rima D. Apple and Janet Golden (eds), *Mothers & Motherhood: Readings in American History* (Columbus, OH: Ohio State University Press, 1997). For women and occupational health, see Harrison, *Not only the 'Dangerous Trades'*; Carolyn Malone, *Women's Bodies and Dangerous Trades in England, 1880–1914* (Woodbridge: Boydell, 2003).

30. Barbara Harrison, 'Women and Health', in June Purvis (ed.), *Women's History: Britain, 1850–1945* (London: UCL Press, 1995), pp. 157–92, quotes on p. 182.
31. John Pickstone, 'Production, Community and Consumption: The Political Economy of Twentieth-Century Medicine', in Roger Cooter and John Pickstone (eds), *Companion to Medicine in the Twentieth Century* (London: Harwood Academic, 2000), pp. 1–20.
32. And has been questioned in Vicky Long and Hilary Marland, 'From Danger and Motherhood to Health and Beauty: Health Advice for the Factory Girl in Early Twentieth-Century Britain', *Twentieth Century British History*, 20 (2009), 454–81.
33. Dorothy Porter, 'The Healthy Body', in Cooter and Pickstone (eds), *Companion to Medicine in the Twentieth Century*, pp. 201–16, on p. 204.
34. See, for example, Helen Jones, *Health and Society in Twentieth-Century Britain* (London and New York: Longman, 1994), pp. 11–17; Jane Lewis, *Women in England 1870–1950: Sexual Divisions and Social Change* (Brighton: Wheatsheaf, 1984), pp. 23–44, and for middle-class women Patricia Branca, *Silent Sisterhood: Middle-Class Women in the Victorian Home* (London: Croom Helm, 1975), Part II.
35. Barbara Brookes, 'Women and Reproduction c.1860-1919', in Jane Lewis (ed.), *Labour and Love: Women's Experience of Home and Family 1850–1940* (Oxford: Basil Blackwell, 1986), pp. 149–71, quote on pp. 149–50. For the extension of sexual knowledge and contraception, see Roy Porter and Lesley Hall (eds), *The Facts of Life: The Creation of Sexual Knowledge in Britain, 1650–1950* (New Haven, CT and London: Yale University Press, 1995); Angus McLaren, *Birth Control in Nineteenth-Century England* (London: Croom Helm, 1978); Hera Cook, *The Long Sexual Revolution: English Women, Sex, and Contraception 1800–1975* (Oxford: Oxford University Press, 2004).
36. For example, Roberts, *A Woman's Place*; Lucinda McCray Beier, *For Their Own Good: The Transformation of English Working-Class Health Culture, 1880–1970* (Columbus, OH: Ohio State University Press, 2008); Kate Fisher, *Birth Control, Sex, and Marriage in Britain, 1918–1960* (Oxford and New York: Oxford University Press, 2006).
37. For example, Ellen Jordan, ' "Making Good Wives and Mothers"? The Transformation of Middle-Class Girls' Education in Nineteenth-Century Britain', *History of Education Quarterly*, 31 (1991), 439–62 and Carol Dyhouse, 'Good Wives and Little Mothers: Social Anxieties and the Schoolgirl's Curriculum, 1890–1920', *Oxford Review of Education*, 3 (1977), 21–35.
38. Carol Dyhouse, *Girls Growing Up in Late Victorian and Edwardian England* (London: Routledge & Kegan Paul, 1981).

39. Iris Dove, 'Sisterhood or Surveillance? The Development of Working Girls' Clubs in London 1880–1939', unpublished University of Greenwich PhD thesis, 1996; M.C. Martin, 'Gender, Religion and Recreation: Flora Lucy Freeman and Female Adolescence 1890–1925', in R. Gilchrest, T. Jeffs and J. Spence (eds), *Drawing on the Past: Studies in the History of Community Youth Work* (Leicester: The National Youth Agency, 2006), pp. 61–77.

40. Ellen Ross, *Love and Toil: Motherhood in Outcast London, 1870–1918* (New York and Oxford: Oxford University Press, 1993), p. 154.

41. George R. Sims, 'At the Front Door', in *idem* (ed.), *Living London*, 1 (London: Cassell, 1901–2), p. 31. Cited in Anna Davin, *Growing Up Poor: Home, School and Street in London 1870–1914* (London: Rivers Oram Press, 1996), p. 88.

42. Penny Tinkler, *Constructing Girlhood: Popular Magazines for Girls Growing Up in England 1920–1950* (London: Taylor & Francis, 1995), pp. 13–20.

43. Ibid., p. 20. Chapter 4 explores the often poor health status of the schoolgirl and Chapter 5 health problems confronting working girls.

44. The relationship between the schoolgirl, health and sport has attracted more attention than other aspects of girls' health. See, for example, Kathleen E. McCrone, *Sport and the Physical Emancipation of English Women 1870–1914* (London: Routledge, 1988); Jennifer Hargreaves, *Sporting Females: Critical Issues in the History and Sociology of Women's Sports* (London: Routledge, 1994), especially Chapter 4; Paul Atkinson, 'Fitness, Feminism and Schooling', in Delamont and Duffin (eds), *The Nineteenth-Century Woman*, pp. 92–133.

45. Henry Maudsley, 'Sex in Mind and in Education', *Fortnightly Review*, new series, 15 (April 1874), 466–83, quote on p. 467. See Elaine Showalter, *The Female Malady: Women, Madness and English Culture, 1830–1980* (London: Virago, 1987; first published New York: Pantheon, 1985), pp. 124–5 and Anne Digby, 'Women's Biological Straitjacket', in Susan Mendus and Jane Rendall (eds), *Sexuality and Subordination: Interdisciplinary Studies of Gender in the Nineteenth Century* (London and New York: Routledge, 1989), pp. 192–220, especially pp. 208–14. See also Louise Michele Newman, *Men's Ideas/Women's Realities: Popular Science, 1870–1915* (New York, etc.: Permagon, 1985), Chapter 2, for debates on women's education in the UK and US.

46. Elizabeth Garrett Anderson, 'Sex in Mind and Education: A Reply', *Fortnightly Review*, new series, 15 (May 1874), 582–94.

47. Showalter, *The Female Malady*, p. 124.

48. Mitchell, *The New Girl*. See also, for girls' fiction and female heroines, Mary Cadogan and Patricia Craig, *You're A Brick Angela! The Girls' Story 1839–1985* (London: Victor Gollancz, 1986); Smith, *Empire in British Girls' Literature and Culture*.

49. E. Nesbit, 'The Girton Girl', *Atalanta*, 8 (1895), 755–9, on p. 755. Cited in Mitchell, *The New Girl*, p. 65. The 'Girton Girl' became a popular topos, linked as she was to the first university-based college for women established in 1869. See ibid, Chapter 3 on college experiences as described in novels and periodicals and Carol Dyhouse, *No Distinction of Sex? Women in British Universities, 1870–1939* (London: UCL Press, 1995). Martha Verbrugge also describes the emergence of new girl types in the US, including the 'Gibson Girl' in the mid-1890s, 'tall, proper, sophisticated, apolitical, and

physically active', who by the 1910s evolved into the athletic 'American Girl': Verbrugge, *Active Bodies*, p. 50.

50. Anon. [E. Lynn Linton], 'The Girl of the Period', *Saturday Review*, 25 (14 March 1868), 339–40. Reprinted in E. Lynn Linton, *The Girl of the Period and Other Social Essays*, vol. 1 (London: Richard Bentley & Son, 1883), p. 6. It also helped define Linton in what became a controversial literary career: Nancy Fix Anderson, 'Linton, Elizabeth [Eliza] Lynn (1822–1898)', *DNB*: [http://0-www.oxforddnb.com.pugwash.lib.warwick.ac.uk/view/article/16742, accessed 27 January 2012]. See also Margaret Beetham, *A Magazine of her Own? Domesticity and Desire in the Woman's Magazine, 1800–1914* (London and New York: Routledge, 1996), pp. 105–6 and Merle Mowbray Bevington, *The Saturday Review 1855–1868: Representative Educated Opinion in Victorian England* (New York: AMS Press, 1966), pp. 110–12, 357–8.

51. Linton, *The Girl of the Period*, p. 2.

52. Ibid., p. 5.

53. An impressive body of research and writing has focused on the feminist and suffrage movements of the late Victorian and Edwardian periods, the emergence of the 'New Woman' and her representation in literature and the press. For the relationship between feminism, eugenics and New Woman scholarship, see Angelique Richardson, *Love and Eugenics in the Late Nineteenth Century* (Oxford: Oxford University Press, 2003) and *idem*, 'The Birth of National Hygiene and Efficiency: Women and Eugenics in Britain and America 1865–1915', in Ann Heilmann and Margaret Beetham (eds), *New Women Hybridities: Femininity, Feminism and International Consumer Culture, 1880–1930* (London and New York: Routledge, 2004), pp. 240–62. For representations of the New Woman in print media, see Angelique Richardson and Chris Willis (eds), *The New Woman in Fiction and in Fact: fin-de-siècle Feminisms* (Basingstoke: Palgrave, 2001); Sally Ledger, *The New Woman: Fiction and Feminism at the fin de siècle* (Manchester and New York: Manchester University Press, 1997); Jane Wood, *Passion and Pathology in Victorian Fiction* (Oxford: Oxford University Press, 2001), Chapter 4; M.E. Tusan, 'Inventing the New Woman: Print Culture and Identity Politics during the Fin-de-Siècle', *Victorian Periodicals Review*, 312 (Summer 1998), 169–82.

54. Mary Wollstonecraft, *A Vindication of the Rights of Woman* (1792); with an introduction by Miriam Brody (London: Penguin, 1985). For nineteenth-century readings of Wollstonecraft, see Nicholas McGuinn, 'George Eliot and Mary Wollstonecraft', in Delamont and Duffin (eds), *The Nineteenth-Century Woman*, pp. 188–205. For discourses on female biology and sex difference, see Thomas Laqueur, *Making Sex: Body and Gender from the Greeks to Freud* (Cambridge, MA and London: Harvard University Press, 1990) and Londa Schiebinger, *The Mind Has No Sex? Women in the Origins of Modern Science* (Cambridge, MA and London: Harvard University Press, 1989).

55. Digby, 'Women's Biological Straitjacket'; Ornella Moscucci, *The Science of Woman: Gynaecology and Gender in England 1800–1929* (Cambridge: Cambridge University Press, 1990), especially Chapter 1.

56. See Hilary Marland, *Dangerous Motherhood: Insanity and Childbirth in Victorian Britain* (Houndmills: Palgrave Macmillan, 2004) for the emergence of

specialists in obstetrics and psychiatry and inter-professional debate and competition. See Anne Digby, *Making a Medical Living: Doctors and Patients in the English Market for Medicine, 1720–1911* (Cambridge: Cambridge University Press, 1994), Chapter 9 for expanding practice opportunities with women and children.

57. A vast literature engages with doctors' pronouncements on women's nature, the impact of their reproductive lives on their health and their susceptibility to illness. See, for example, Marland, *Dangerous Motherhood*; Moscucci, *The Science of Woman*; Cynthia Eagle Russett, *Sexual Science: The Victorian Construction of Womanhood* (Cambridge, MA: Harvard University Press, 1989), and for the link between medicine and literary interpretations, Wood, *Passion and Pathology*, especially Chapter 1 on the medicalisation of womanhood. For North America, see, for example, Wendy Mitchinson, *The Nature of their Bodies: Women and their Doctors in Victorian Canada* (Toronto, ON: University of Toronto Press, 1991) and Nancy M. Theriot, 'Negotiating Illness: Doctors, Patients, and Families in the Nineteenth Century', *Journal of the History of the Behavioral Sciences*, 37 (2001), 349–68, which explores the impact of women patients on medical diagnoses and interpretations of illness.

58. Digby, 'Women's Biological Straitjacket', p. 208. See Kate Flint, *The Woman Reader 1837–1914* (Oxford: Clarendon, 1993), Chapter 4, for writing and reading on medical themes in the Victorian and Edwardian period.

59. Clement Dukes, 'Health at School', in Malcolm Morris (ed.), *The Book of Health* (London, Paris and New York: Cassell, 1883), pp. 677–725, quote on p. 724. The same volume included a chapter by J. Crichton-Browne on 'Education and the Nervous System', which reinforced Maudsley's arguments.

60. Cf. Christopher Lawrence, for example, who formulated the idea that modern medicine was tied up with the emergence of medical elites with strong ties to medical science and institutional practice: *Medicine in the Making of Modern Britain 1700–1920* (London and New York: Routledge, 1994), Chapter 3.

61. Ina Zweiniger-Bargielowska, 'Raising a Nation of "Good Animals": The New Health Society and Health Education Campaigns in Interwar Britain', *Social History of Medicine*, 20 (2007), 73–89; Bruce Haley, *The Healthy Body and Victorian Culture* (Cambridge, MA and London: Harvard University Press, 1978).

1 Unstable Adolescence: Medicine and the 'Perils of Puberty' in Late Victorian and Edwardian Britain

1. For the US case, see Joan Jacobs Brumberg, *The Body Project: An Intimate History of American Girls* (New York: Vintage, 1998) and Margaret A. Lowe, *Looking Good: College Women and Body Image, 1875–1930* (Baltimore, MD and London: Johns Hopkins University Press, 2003); for young women in post-First World War Denmark, Birgitte Søland, *Becoming Modern: Young Women and the Reconstruction of Womanhood in the 1920s* (Princeton, NJ and Oxford: Princeton University Press, 2000), especially Chapter 2.

2. Jules Michelet, *L'Amour* (Paris: L. Hachette & Cie, 1858), p. 48. Michelet's work had a huge readership in the US and Britain. See also Patricia A. Vertinsky, *The Eternally Wounded Woman: Women, Doctors, and Exercise in the Late Nineteenth Century* (Urbana and Chicago, IL: University of Illinois Press, 1994; first published Manchester: Manchester University Press, 1989), Chapter 1.

3. Vertinsky, *The Eternally Wounded Woman*, p. 39.

4. Mary Lynn Stewart, *For Health and Beauty: Physical Culture for Frenchwomen, 1880s-1930s* (Baltimore, MD and London: Johns Hopkins University Press, 2001), p. 195.

5. See Kate Flint, *The Woman Reader 1837–1914* (Oxford: Clarendon, 1993) for the publication of doctors in lay media and female reading practices on health matters.

6. Carol Dyhouse, *Girls Growing Up in Late Victorian and Edwardian England* (London: Routledge & Kegan Paul, 1981), p. 116.

7. For the dominance of boys in these accounts, see, for example, John Gillis, *Youth and History: Tradition and Change in European Age Relations 1770-Present* (New York and London: Academic Press, 1974); John Springhall, *Youth, Empire and Society: British Youth Movements 1883–1940* (London: Croom Helm, 1977); Harry Hendrick, *Images of Youth: Age, Class, and the Male Youth Problem, 1880–1920* (Oxford: Clarendon Press, 1990).

8. Vertinsky, *The Eternally Wounded Woman*, p. 46.

9. Anne Digby, 'Women's Biological Straitjacket', in Susan Mendes and Jane Rendall (eds), *Sexuality and Subordination: Interdisciplinary Studies of Gender in the Nineteenth Century* (London and New York: Routledge, 1989), pp. 192–220. See also Hilary Marland, *Dangerous Motherhood: Insanity and Childbirth in Victorian Britain* (Houndmills: Palgrave Macmillan, 2004), Chapter 2 for contestation of the management of women's reproductive health.

10. See, for example, Carroll Smith-Rosenberg and Charles Rosenberg, 'The Female Animal: Medical and Biological Views of Woman and Her Role in Nineteenth-Century America'; Ann Douglas Wood, ' "The Fashionable Diseases": Women's Complaints and Their Treatment in Nineteenth-Century America'; and Regina Markell Moranz, 'The Perils of Feminist History', in Judith Walzer Leavitt (ed.), *Women and Health in America* (Madison, WI: University of Wisconsin Press, 1984), pp. 12–27, 222–38, 239–45; Cynthia Eagle Russett, *Sexual Science: The Victorian Construction of Womanhood* (Cambridge, MA and London: Harvard University Press, 1989); Ornella Moscucci, *The Science of Woman: Gynaecology and Gender in England 1800–1929* (Cambridge: Cambridge University Press, 1990), Chapter 1; Wendy Mitchinson, *The Nature of their Bodies: Women and their Doctors in Victorian Canada* (Toronto, ON: University of Toronto Press, 1991).

11. Anne Digby, *Making a Medical Living: Doctors and Patients in the English Market for Medicine, 1720–1911* (Cambridge: Cambridge University Press, 1994), Chapter 9.

12. Moscucci, *The Science of Woman*, p. 104.

13. Vertinsky, *The Eternally Wounded Woman*, p. 47; Herbert Spencer, *Education: Intellectual, Moral and Physical* (London: G. Manwaring, 1861), pp. 179–80, 186, 188. Burstyn has proposed a link between Spencer's thesis and the

principle of the conservation of energy introduced into physics in the 1840s: Joan N. Burstyn, 'Education and Sex: The Medical Case against Higher Education for Women in England, 1870–1900', *Proceedings of the American Philosophical Society*, 117 (1973), 79–89, on p. 85.

14. Elaine Showalter and English Showalter, 'Victorian Women and Menstruation', in Martha Vicinus (ed.), *Suffer and be Still: Women in the Victorian Age* (London: Methuen, 1980; first published Bloomington, IN: Indiana University Press, 1972), pp. 38–44, quote on p. 43.

15. E. Clarke, *Sex in Education: Or, a Fair Chance for Girls* (Boston, MA: James R. Osgood, 1874), p. 18.

16. Ibid., p. 39.

17. Ibid., p. 19.

18. Julie-Marie Strange, 'The Assault on Ignorance: Teaching Menstrual Etiquette in England, c.1920s to 1960s', *Social History of Medicine*, 14 (2001), 247–65, on p. 253.

19. Henry Maudsley, 'Sex in Mind and in Education', *Fortnightly Review*, new series, 15 (April 1874), 466–83, quotes on pp. 468, 475–6. See T.H. Turner, 'Henry Maudsley (1835–1918)', *Oxford Dictionary of National Biography* (hereafter *DNB*): [http://www.oxforddnb.com.pugwash.lib.warwick.ac.uk/view/article/37747, accessed 17 March 2012]; Andrew Scull, Charlotte MacKenzie and Nicholas Hervey, *Masters of Bedlam: The Transformation of the Mad-Doctoring Trade* (Princeton, NJ: Princeton University Press, 1996), chapter 8, 'Degeneration and Despair: Henry Maudsley (1835–1918)'.

20. Flint, *The Woman Reader*, p. 57.

21. Elizabeth Garrett Anderson, 'Sex in Mind and Education: A Reply', *Fortnightly Review*, new series, 15 (May 1874), 582–94, quote on p. 590.

22. Claire Brock, 'Surgical Controversy at the New Hospital for Women, 1872–1892', *Social History of Medicine*, 24 (2011), 608–23; Jo Manton, *Elizabeth Garrett Anderson* (London: Methuen, 1965); M.A. Elston, 'Anderson, Elizabeth Garrett (1836–1917)', *DNB*: [http://www.oxforddnb.com.pugwash.lib.warwick.ac.uk/view/article/30406, accessed 17 March 2012]. See Regina Markell Mortanz and Sue Zschoche, 'Professionalism, Feminism, and Gender Roles: A Comparative Study of Nineteenth-Century Medical Therapeutics', in Leavitt (ed.), *Women and Health in America*, pp. 406–21, for a comparison of the therapeutic approaches of male and female doctors in two Boston hospitals and Moscucci, *The Science of Woman*, for controversies between radical and conservative surgeons on gynaecological interventions, especially Chapters 4 and 5.

23. 'British Medical Association. Fifty-Fourth Annual Meeting, Held at Brighton, 1886', Report on the Presidential Address, *Lancet*, 2 (14 August 1886), 313–16, on p. 314.

24. Ibid.; William Withers Moore, *Lancet* excerpt, quoted by Emily Pfeiffer, *Women and Work: An Essay Treating on the Relation to Health and Physical Development, of the Higher Education of Girls, and the Intellectual or More Systematised Effort of Women* (London: Trübner & Co., 1888), p. 67.

25. T.S. Clouston, *Clinical Lectures on Mental Diseases*, 3rd edn (London: J. and A. Churchill, 1892), p. 568.

26. J. Mortimer-Granville, *Youth: Its Care and Culture. An Outline of Principles for Parents and Guardians* (London: David Bogue, 1880). Mortimer-Granville claimed to have invented the electric vibrator to treat nerve disorders, though asserted that he had never employed this to treat women: Rachel P. Maines, *The Technology of Orgasm: 'Hysteria,' the Vibrator, Women's Sexual Satisfaction* (Baltimore, MD and London: Johns Hopkins University Press, 1999), pp. 93–4, 98–9.
27. Mortimer-Granville, *Youth*, pp. 74, 79, 89–90, 95.
28. Ibid., p. 88.
29. Ibid., p. 126.
30. Ibid., p. 37.
31. Both E.J. Tilt and George Black referred to the relationship of menstruation with urbanisation: E.J Tilt, *On the Preservation of the Health of Women at the Critical Periods of Life* (London: John Churchill, 1851), pp. 36–7 and George Black, *The Young Wife's Advice Book: A Guide for Mothers on Health and Self-Management*, 6th edn (London: Ward, Lock and Co., 1888), p. 5.
32. See also the extracts in Pat Jalland and John Hooper (eds), *Women from Birth to Death: The Female Life Cycle in Britain 1830–1914* (Brighton: Harvester, 1986), Part 2 'Menstruation and Adolescence' and Deborah Gorham, *The Victorian Girl and the Feminine Ideal* (London and Canberra: Croom Helm, 1982), Chapter 5, for advice on the management of puberty. Julie-Marie Strange has explored menstrual hygiene and views on mental disability in 'The Assault on Ignorance'. See also Showalter and Showalter, 'Victorian Women and Menstruation'.
33. Tilt, *On the Preservation of the Health*, p. 31.
34. Ibid., p. 39.
35. Black, *The Young Wife's Advice Book*, p. 5. See Flint, *The Woman Reader*, pp. 57–8 for advice on the avoidance of stimuli amongst young women.
36. Black, *The Young Wife's Advice Book*, p. 6.
37. Tilt, *On the Preservation of the Health*, p. 42; *idem, Elements of Health, and Principles of Female Hygiene* (London: Henry G. Bohn, 1852), pp. 141, 150–6.
38. Garrett Anderson, 'Sex in Mind and Education: A Reply', p. 590.
39. Anon., *The Ladies' Physician: A Guide for Women in the Treatment of Their Ailments: By a London Physician*, 9th edn (London, Paris & Melbourne: Cassell, 1891), p. 1.
40. Robert Reid Rentoul, *Race Culture; Or, Race Suicide? (A Plea for the Unborn)* (London and Felling-on-Tyne: Walter Scott Publishing, 1906). See Frank Mort, *Dangerous Sexualities: Medico-Moral Politics in England since 1830* (London: Routledge & Kegan Paul, 1987), pp. 171, 182–3 for responses to Rentoul's views.
41. Robert Reid Rentoul, *The Dignity of Woman's Health and the Nemesis of its Neglect (A Pamphlet for Women and Girls)* (London: J. & A. Churchill, 1890), p. xxviii.
42. Ibid., p. xvi.
43. Ibid., pp. xxx, xxviii.
44. Ibid., pp. 45–6.
45. Rentoul cited the 50th Annual Report of the Registrar-General, and the high numbers of deaths from ovarian disease, disease of the uterus, menstrual

disorders and accidents of childbirth, which he attributed to lack of care during puberty: ibid., p. xvi.

46. John Thorburn, *Female Education from a Medical Point of View* (Manchester: J.E. Cornish, 1884), pp. 2, 4, 15.

47. Henry Maudsley, *The Physiology and Pathology of Mind: A Study of Its Distempers, Deformities and Disorders*, 2nd edn (London: Macmillan, 1868), p. 341.

48. Irvine Loudon, 'Chlorosis, Anaemia, and Anorexia Nervosa', *British Medical Journal*, 281 (20–27 December 1980), 1669–75, p. 1671. Loudon has suggested that chlorosis and anorexia nervosa were closely related, and represented 'manifestation of the same type of psychological reaction to the turbulence of puberty and adolescence' (p. 1675). For an account of chlorosis from the classical period to the early twentieth century, see Helen King, *The Disease of Virgins: Green Sickness, Chlorosis and the Problems of Puberty* (London and New York: Routledge, 2004). See also Karl Figlio, 'Chlorosis and Chronic Disease in Nineteenth-Century Britain: The Social Construction of Somatic Illness in a Capitalist Society', *Social History*, 3 (1978), 167–97 and Joan Jacobs Brumberg, 'Chlorotic Girls, 1870–1920: A Historical Perspective on Female Adolescence', in Leavitt (ed.), *Women and Health in America*, pp. 186–95.

49. Sir Andrew Clark, 'Anaemia or Chlorosis of Girls, Occurring more Commonly between the Advent of Menstruation and the Consummation of Womanhood', *Lancet*, 2 (19 November 1887), 1003–5, quote on p. 1004.

50. Ibid., p. 1005.

51. For the thesis that the absence of sexual activity and the 'empty womb' prompted green sickness and hysteria, see Laurinda S. Dixon, *Perilous Chastity: Women and Illness in Pre-Enlightenment Art and Medicine* (Ithaca, NY and London: Cornell University Press, 1995).

52. A.L. Galabin, *Diseases of Women*, 5th edn (London: J. & A. Churchill, 1893), pp. 473, 475–6.

53. John Thorburn, *A Practical Treatise on the Diseases of Women* (London: Charles Griffin, 1885), pp. 104–5.

54. Edward Johnson, *The Hydropathic Treatment of Diseases Peculiar to Women; and of Women in Childbed; with some Observations on the Management of Infants* (London: Simpkin, Marshall, and Co., 1850), pp. 39–40.

55. Ibid., pp. 44–6, quotes on p. 44.

56. E.H. Ruddock, *The Lady's Manual of Homoeopathic Treatment in the Various Derangements Incident to Her Sex*, 9th edn (London: The Homoeopathic Publishing Company, 1886), p. 23.

57. Ibid., p. 27.

58. Ibid., pp. 27, 29–30.

59. Ibid., p. 29.

60. There is a vast literature on hysteria, including Sander Gilman, Helen King, Roy Porter, George S. Rousseau and Elaine Showalter, *Hysteria beyond Freud* (Berkeley, CA: University of California Press, 1993); Carroll Smith-Rosenberg, 'The Hysterical Woman: Sex Roles and Role Conflict in 19th-Century America', *Social Research*, 39 (1979), 652–78; Elaine Showalter, *The Female Malady: Women, Madness and English Culture, 1830–1980* (London: Virago, 1987; first published New York: Pantheon, 1985), Chapter 6. See also

Flint, *The Woman Reader*, pp. 57–60 for the relationship between hysteria and reading, and for further examples of medical writing on the condition, Jalland and Hooper (eds), *Women from Birth to Death*, Section 2.4.

61. Walter Johnson, *The Morbid Emotions of Women; Their Origin, Tendencies, and Treatment* (London: Simpkin, Marshall, and Co., 1850), pp. 227, 233, 232.

62. For examples of cases affecting young women, see J.S. Bristowe, 'Hysteria and Its Counterfeit Presentments', *Lancet*, 1 (13 June 1885), 1069–72, 1114–17 and for a discussion of causality, Thomas D. Savill, 'The Psychology and Psychogeniesis of Hysteria and the Rôle of the Sympathetic System', *Lancet*, 1 (13 February 1909), 443–8.

63. W.S. Playfair, *The Systematic Treatment of Nerve Prostration and Hysteria* (London: Smith Elder, & Co., 1883), p. 81.

64. F.C. Skey, *Hysteria: Six Lectures Delivered to the Students of St. Bartholomew's Hospital, 1866* (London: Longmans, Green, Reader & Dyer, 1867), pp. 54–5.

65. Ibid., p. 55.

66. H.B. Donkin, 'Hysteria', in D. Hack Tuke (ed.), *A Dictionary of Psychological Medicine* (London: J. & A. Churchill, 1892), pp. 618–27, quote on p. 620.

67. Ibid. See also Sally Ledger and Roger Luckhurst (eds), *The fin de siècle: A Reader in Cultural History, c.1880–1900* (Oxford: Oxford University Press, 2000), p. 245.

68. Henry Maudsley, *The Pathology of Mind: A Study of its Distempers, Deformities and Disorders*, 3rd edn of 2nd part of *The Physiology and Pathology of Mind* (London: Macmillan, 1879), p. 450.

69. Ibid., 1895 edn (London: Macmillan, 1895), p. 397.

70. W.S. Playfair, 'The Systematic Treatment of Functional Neurosis', in Tuke, *A Dictionary of Psychological Medicine*, vol. II, pp. 850–7, quote on p. 854.

71. Showalter, *The Female Malady*, p. 134. For neurasthenia, see, for example, Janet Oppenheim, *'Shattered Nerves': Doctors, Patients, and Depression in Victorian England* (Oxford: Oxford University Press, 1991); Marijke Gijswijt-Hofstra and Roy Porter (eds), *Cultures of Neurasthenia From Beard to the First World War* (Amsterdam and New York: Rodopi, 2001), including on Playfair, Chandak Sengoopta, ' "A Mob of Incoherent Symptoms"? Neurasthenia in British Medical Discourse, 1860–1920' and Hilary Marland, ' "Uterine mischief": W.S. Playfair and his Neurasthenic Patients', pp. 97–115, 117–39.

72. Playfair, 'The Systematic Treatment of Functional Neurosis', pp. 850–7, quote on p. 851.

73. Jane Wood, *Passion and Pathology in Victorian Fiction* (Oxford: Oxford University Press, 2001), Chapter 4; Gail Cunningham, *The New Woman and the Victorian Novel* (London and Basingstoke: Macmillan, 1978).

74. For anorexia nervosa, see Hilde Bruche, *Eating Disorders: Obesity, Anorexia Nervosa, and the Person Within* (New York: Basic Books, 1973); Joan Jacobs Brumberg, *Fasting Girls: The History of Anorexia Nervosa* (Cambridge, MA: Harvard University Press, 1988).

75. Brumberg, *Fasting Girls*, p. 129; William Gull, 'Anorexia Nervosa (Apepsia Hysterica, Anorexia Hysteria)', *Transactions of the Clinical Society of London*, 7 (1874), 22–8, on p. 22.

76. Gull, 'Anorexia Nervosa'. See Brumberg, *Fasting Girls*, pp. 118–25, for a full account of this influential paper.

77. Gull, 'Anorexia Nervosa', pp. 26, 27.
78. Brumberg, *Fasting Girls*, pp. 122–3.
79. William Gull, 'Anorexia Nervosa', *Lancet*, 1 (17 March 1888), 516–17, quote on p. 517.
80. W.S. Playfair, 'Note on the So-Called "Anorexia Nervosa"', *Lancet*, 1 (28 April 1888), 817–18, quote on p. 818.
81. Sarah Grand, *The Heavenly Twins* (London: George Heinemann, 1893), p. 594. Cited in Showalter, *The Female Malady*, p. 137. See also Sally Ledger, *The New Woman: Fiction and Feminism at the fin de siècle* (Manchester and New York: Manchester University Press, 1997).
82. Jalland and Hooper (eds.), *Women from Birth to Death*, p. 59.
83. R. Clement Lucas, 'Two Cases of "Corset Cancer" with Remarks on the Origin and Spread of Cancer', *Lancet*, 1 (2 April 1904), 929–31.
84. E.B. Duffey, *What Women Should Know. A Woman's Book about Women. Containing Practical Information for Wives and Mothers* (Philadelphia, PA: J.M. Stoddart [1873]), p. 40.
85. Ibid., p. 41.
86. Howard A. Kelly, *Medical Gynecology* (New York and London: Appleton & Co., 1908), p. 70; Thorburn, *A Practical Treatise on the Diseases of Women*, p. 98.
87. Frederick Treves, 'The Influence of Dress on Health', in Malcolm Morris (ed.), *The Book of Health* (London, Paris and New York: Cassell, 1883), pp. 461–517, on pp. 500–6.
88. Ibid., p. 505.
89. See Chapter 2 for health advice on dress and for detailed studies of the corset, Valerie Steel, *The Corset: A Cultural History* (New Haven, CT and London: Yale University Press, 2001); Leigh Summers, *Bound to Please: A History of the Victorian Corset* (Oxford: Berg, 2001); Susan Aspinall, 'Nurture as Well as Nature: Environmentalism in Representations of Women and Exercise in Britain from the 1880s to the Early 1920s', unpublished University of Warwick, PhD thesis, 2008, Chapter 3.
90. Jalland and Hooper (eds), *Women from Birth to Death*, pp. 216–17.
91. Robert Lawson Tait, *The Diseases of Women*, 2nd edn (New York: William Wood & Co., 1879), p. 30.
92. For the Baker Brown controversy, see Jalland and Hooper (eds), *Women from Birth to Death*, pp. 217–18, 250–65. Baker Brown was a renowned surgeon and in 1865 elected President of the Medical Society of London. He carried out numerous clitoridectomies at his London Surgical Home, and was expelled from the Obstetrical Society of London in 1867 for failing to obtain the consent of his patients. He had many supporters but also numerous detractors.
93. Wood, *Passion and Pathology*, p. 21.
94. H. MacNaughton-Jones, 'The Relation of Puberty and the Menopause to Neurasthenia', *Lancet*, 1 (29 March 1913), 879–81, quote on p. 880.
95. Kelly, *Medical Gynecology*, pp. 72, 291.
96. Ibid., pp. 297–8.
97. For the relationship between children, health and medicine in the late nineteenth and twentieth centuries, see the essays in Roger Cooter (ed.), *In the Name of the Child: Health and Welfare 1880–1940* (London and

New York: Routledge, 1992) and Marijke Gijswift-Hofstra and Hilary Marland (eds), *Cultures of Child Health in Britain and the Netherlands in the Twentieth Century* (Amsterdam and New York: Rodopi, 2003), and for a broad overview of research on childhood and adolescence, Alysa Levene, 'Childhood and Adolescence', in Mark Jackson (ed.), *The Oxford Handbook of the History of Medicine* (Oxford: Oxford University Press, 2011), pp. 321–37.

98. Vertinsky, *The Eternally Wounded Woman*, p. 171.

99. See R.A. Soloway, *Demography and Degeneration: Eugenics and the Declining Birthrate in Twentieth Century Britain* (Chapel Hill, NC: University of North Carolina Press, 1990); Nancy Stepan, *The Idea of Race in Science: Great Britain, 1800–1960* (London: Palgrave Macmillan, 1982); for national fitness and citizenship amongst children and youth, see, for example, Dyhouse, *Girls Growing Up*; John Welshman, 'Child Health, National Fitness, and Physical Education in Britain, 1900–1940' and Bernard Harris, 'Educational Reform, Citizenship and the Origins of the School Medical Service', in Gijswijt-Hofstra and Marland (eds), *Cultures of Child Health*, pp. 61–84, 85–101 and the introduction to this volume, pp. 7–30; for motherhood, the infant welfare movement, war and citizenship, Anna Davin, 'Imperialism and Motherhood', *History Workshop Journal*, 5 (1978), 9–66; Deborah Dwork, *War is Good for Babies and Other Young Children: A History of the Infant and Child Welfare Movement in England 1898–1918* (London and New York: Tavistock, 1987); Lara Marks, 'Mothers, Babies and Hospitals: "The London" and the Provision of Maternity Care in East London', in Valerie Fildes, Lara Marks and Hilary Marland (eds), *Women and Children First: International Maternal and Infant Welfare 1870–1945* (London and New York: Routledge, 1992), pp. 48–73.

100. Sir William Taylor, 'An Address on the Medical Profession in Relation to the Army', *Lancet*, 2 (18 October 1902), 1088–91. Cited in Lucy Bland and Lesley A. Hall, 'Eugenics in Britain: The View from the Metropole', in Alison Bashford and Phillipa Levine (eds), *The Oxford Handbook of the History of Eugenics* (Oxford: Oxford University Press, 2010), pp. 213–27, on p. 213.

101. G. Stanley Hall, *Adolescence: Its Psychology and the Relation to Physiology, Anthropology, Sociology, Sex, Crime, Religion and Education* (New York: D. Appleton, 1904).

102. Dyhouse, *Girls Growing Up*, p. 122.

103. Ibid., pp. 122–9.

104. G. Stanley Hall, 'The Budding Girl', in *idem*, *Educational Problems*, vol. II (New York and London: D. Appleton, 1911), p. 33.

105. Ibid., pp. 1, 2–17.

106. Ibid., p. 34.

107. G. Stanley Hall, *Youth: Its Education, Regimen, and Hygiene* (New York: D. Appleton, 1906), pp. 309–14.

108. Ibid., p. 314.

109. Phyllis Blanchard, *The Care of the Adolescent Girl: A Book for Teachers, Parents, and Guardians*, with prefaces by Mary Scharlieb and G. Stanley Hall (London: K. Paul, Trench, Trübner, 1921), p. 115. Dementia praecox refers to a chronic, deteriorating psychotic disorder usually beginning in the late teens or early adulthood, defined by German psychiatrist Emil Kraeplin after 1893; it would eventually be reframed and relabelled as schizophrenia.

110. M.E. Findlay, 'The Education of Girls', *The Paidologist*, VII (1905), 83–93, quote on p. 86.
111. Arabella Kenealy, *Feminism and Sex-Extinction* (London: T. Fisher Unwin, [1920]), p. 110. Kenealy's views on female athletes will be explored in Chapter 3.
112. J.W. Slaughter, *The Adolescent* (London: G. Allen, 1907), p. 95.
113. T.N. Kelynack, *Youth* (London: Charles H. Kelly, 1918), p. 34.
114. Ibid., p. 35.
115. Kelynack was also editor of the journal *The Child* (1910–27), and collated and made accessible medical information on topics such as tuberculosis and alcoholism. See Linda Bryder, 'Kelynack, Theophilus Nicholas (1866–1944)', *DNB*: [http://0-www.oxforddnb.com.pugwash.lib.warwick.ac.uk/view/article/51798, accessed 22 January 2012].
116. Bland and Hall, 'Eugenics in Britain: The View from the Metropole', p. 214. For eugenics in Britain, see Pauline M.H. Mazumdar, *Eugenics, Human Genetics and Human Failings: The Eugenics Society, Its Sources and Its Critics in Britain* (London: Routledge, 1992); G.R. Searle, *Eugenics and Politics in Britain, 1900–14* (Leyden: Noordhoff, 1976), *idem*, *The Quest for National Efficiency: A Study in British Politics and Political Thought, 1899–1914* (Oxford: Blackwell, 1971).
117. Greta Jones, 'Women and Eugenics in Britain: The Case of Mary Scharlieb, Elizabeth Sloan Chesser, and Stella Browne', *Annals of Science*, 51 (1995), 481–502; Greta Jones, 'Chesser [*née* Sloan], Elizabeth Macfarlane (1877–1940)', *DNB*: [http://0-www.oxforddnb.com.pugwash.lib.warwick.ac.uk/view/article/57661, accessed 22 January 2012]; Greta Jones, 'Scharlieb [*née* Bird], Dame Mary Ann Dacomb (1845–1930)', *DNB*: [http://0-www.oxforddnb.com.pugwash.lib.warwick.ac.uk/view/article/35968, accessed 22 January 2012].
118. Dorothy Porter, ' "Enemies of the Race": Biologism, Environmentalism, and Public Health in Edwardian England', *Victorian Studies*, 35 (1991), 159–78. See for debates on the impact of eugenics in shaping policy and health matters in the Edwardian period, Searle, *Eugenics and Politics*; Greta Jones, *Social Hygiene in Twentieth-Century Britain* (London: Croom Helm, 1986).
119. Jones, 'Women and Eugenics in Britain', pp. 485–6; Mary Ann Elston, ' "Run by Women, (mainly) for Women": Medical Women's Hospitals in Britain, 1866–1948', in Anne Hardy and Lawrence Conrad (eds), *Women and Modern Medicine* (Amsterdam and New York: Rodopi, 2001), pp. 73–107, on p. 81. For Scharlieb and birth control, see Lesley A. Hall, 'A Suitable Job for a Woman: Women Doctors and Birth Control to the Inception of the NHS', in ibid., pp. 127–47.
120. Mary Scharlieb, 'Adolescent Girlhood under Modern Conditions, with Special Reference to Motherhood', *Eugenics Review*, 1 (1909), 174–83, on p. 177.
121. Ibid., p. 179.
122. Mary A.D. Scharlieb, 'Adolescent Girls from the View-Point of the Physician', *The Child*, 1: 12 (September 1911), 1013–31, on p. 1015.
123. Ibid., p. 1030.
124. Jones, 'Women and Eugenics in Britain', p. 487.

125. W.S. Playfair, 'The Nervous System in Relation to Gynaecology', in Thomas Clifford Allbutt and W.S. Playfair (eds), *A System of Gynaecology by Many Writers* (London: Macmillan, 1896), pp. 220–31, on pp. 220–2.

126. Ibid., p. 221; Anon., 'The New Woman-Old Style', *The Speaker* (12 January 1895), 39–41, quote on p. 40.

127. These relationships will be explored in Chapter 4 in connection with the health of schoolgirls, which directly engaged physicians in practical work with young women, providing them with a great deal of evidence to feed into their publications.

128. Nancy M. Theriot, 'Negotiating Illness: Doctors, Patients, and Families in the Nineteenth Century', *Journal of the History of the Behavioral Sciences*, 37 (2001), 349–68.

129. See also Smith-Rosenberg, 'The Hysterical Woman' which argues that doctors could act as mediators as well as oppressors of their female patients in the case of hysteria, and Marland, *Dangerous Motherhood*, especially Chapter 3 for women's self-diagnosis in cases of puerperal insanity.

130. A.B. Barnard, *The Girl's Book About Herself* (London: Cassell, 1912), p. 10. For the expression of similar concerns in Denmark, see Søland, *Becoming Modern*, Chapter 2.

131. Michel Foucault, *The History of Sexuality: Volume 1, An Introduction* (New York: Vintage Books, 1980).

2 Reinventing the Victorian Girl: Health Advice for Girls in the Late Nineteenth and Early Twentieth Centuries

1. F.H. [Frank Hird], 'Women's Work: Its Value and Possibilities', *Girl's Own Paper* (*GOP*), XVI: 774 (27 October 1894), 51.

2. Anon., 'Girls Physical Training', *Health* (8 November 1895), 85.

3. Anon., 'Exercise for Women', *The Woman's Signal*, 6: 139 (27 August 1896), 136. Gerritsen Collection: http://gerritsen.chadwyck.com.

4. See, for this process in broader perspective, Helen Jones, *Health and Society in Twentieth-Century Britain* (London and New York: Longman, 1994), especially Chapter 2.

5. Bruce Haley, *The Healthy Body and Victorian Culture* (Cambridge, MA and London: Harvard University Press, 1978).

6. Lori Anne Loeb, *Consuming Angels: Advertising and Victorian Women* (New York and Oxford: Oxford University Press, 1994).

7. For the proliferation of domestic medical guides in the nineteenth century, see Hilary Marland, ' "The Diffusion of Useful Knowledge": Household Practice, Domestic Medical Guides and Medical Pluralism in Nineteenth-Century Britain', in Robert Jütte (ed.), *Medical Pluralism: Past and Present, Medizin, Gesellschaft und Geschichte Beiheft* (2013), pp. 81–100.

8. Carol Zisowitz Stearns and Peter N. Stearns, *Anger: The Struggle for Emotional Control in America's History* (Chicago and London: University of Chicago Press, 1986), pp. 36, 53.

9. Maud Curwen and Ethel Herbert, *Simple Health Rules and Health Exercises for Busy Women and Girls* (London: Simpkin, Marshall, Hamilton, Kent & Co., 1912), pp. 1–2.

10. Florence Harvey Richards, *Hygiene for Girls: Individual and Community* (London: D.C. Health, 1913).
11. Anon. [E. Lynn Linton], 'The Girl of the Period', *Saturday Review*, 25 (14 March 1868), 339–40. Linton dedicated her novel, *The One Too Many* (1894) to 'THE SWEET GIRLS STILL LEFT AMONG US', who were content to be dutiful, innocent and sheltered: Kate Flint, *The Woman Reader 1837–1914* (Oxford: Clarendon, 1993), p. 305.
12. Mary Whitley, *Every Girl's Book of Sport, Occupation and Pastime* (London: George Routledge and Sons, 1897), p. 1.
13. Jennie Chandler, 'Hygiene for Women: A New Education for Women', *Good Health*, IV: 91 (30 June 1894), 187.
14. Anon., 'The New Woman', *Good Health*, VI: 131 (6 April 1895), 15.
15. Your Friend, The Editor, 'Chat with the Girl of the Period', *The Girl's Realm Annual* (November 1898–October 1899) (London: Hutchinson, 1899), p. 216.
16. Carol Dyhouse, *Girls Growing Up in Late Victorian and Edwardian England* (London: Routledge & Kegan Paul, 1981), p. 118.
17. Margaret Beetham, *A Magazine of her Own? Domesticity and Desire in the Woman's Magazine, 1800–1914* (London and New York: Routledge, 1996), p. 138.
18. For the role of individual responsibility, marked, for example, by campaigns to reduce levels of infant mortality and to fight tuberculosis, see Christopher Lawrence, *Medicine in the Making of Modern Britain 1700–1920* (London and New York: Routledge, 1994), especially p. 74; Michael Worboys, 'The Sanatorium Treatment for Consumption in Britain, 1890–1914', in John V. Pickstone (ed.), *Medical Innovations in Historical Perspective* (London: Macmillan, 1992), pp. 47–71; Jane Lewis, 'Providers, "Consumers", the State and the Delivery of Health-Care Services in Twentieth-Century Britain', in Andrew Wear (ed.), *Medicine in Society* (Cambridge: Cambridge University Press, 1992), pp. 317–46.
19. Ina Zweiniger-Bargielowska, *Managing the Body: Beauty, Health, and Fitness in Britain, 1880–1939* (Oxford: Oxford University Press, 2010), p. 121.
20. See Beetham, *A Magazine of her Own?*; Kirsten Drotner, *English Children and their Magazines, 1751–1945* (New Haven, CT and London: Yale University Press, 1988); Penny Tinkler, *Constructing Girlhood: Popular Magazines for Girls Growing Up in England 1920–1950* (London: Taylor & Francis, 1995); Sally Mitchell, *The New Girl: Girls' Culture in England 1880–1915* (New York: Columbia University Press, 1995); Mary Cadogan and Patricia Craig, *You're a Brick Angela! The Girls' Story 1839–1985* (London: Victor Gollancz, 1986).
21. Sally Ledger, *The New Woman: Fiction and Feminism at the fin de siècle* (Manchester and New York: Manchester University Press, 1997); Sally Ledger and Roger Luckhurst (eds), *The fin de siècle: A Reader in Cultural History, c.1880–1900* (Oxford: Oxford University Press, 2000), '4. The New Woman'; Gail Cunningham, *The New Woman and the Victorian Novel* (London and Basingstoke: Macmillan, 1978).
22. Beetham, *A Magazine of Her Own?*, chapter 8, pp. 137–8.
23. Terri Doughty, *Selections from The Girl's Own Paper, 1880–1907* (Peterborough, ON: Broadview, 2004), 'Introduction', p. 7. See Beetham, *A Magazine of Her*

Own?, especially chapter 8; Drotner, *English Children and their Magazines*, section IV.

24. Adrian Bingham, *Gender, Modernity, and the Popular Press in Inter-War Britain* (Oxford: Clarendon, 2004), p. 145.
25. Ibid., pp. 26–9. See also pp. 145–81 for fashion, the female body and sexual morality in the press from the First World War onwards.
26. Thomas Richards, *The Commodity Culture of Victorian England: Advertising and Spectacle 1851–1914* (Stanford, CA: Stanford University Press, 1990), p. 244. Beetham refers to the pages of *Home Chat*, which in 1896 contained advertisements for products promoting multiple benefits, many making contradictory claims and presenting different images of the perfect female body: Beetham, *A Magazine of Her Own?*, pp. 196–7.
27. See Loeb, *Consuming Angels* for the expanding consumption of health, beauty and hygiene products by 'empowered' women in the late nineteenth century, as well as the way their images were used in advertisements for patent medicines and health products. For commodity culture and health more broadly, see Takahiro Ueyama, *Health in the Marketplace: Professionalism, Therapeutic Desires, and Medical Commodification in Late-Victorian London* (Palo Alto, CA: Society for the Promotion of Science and Scholarship, 2010).
28. See Richards, *The Commodity Culture of Victorian England*, especially pp. 228–34.
29. Wilson was the author of several health guides, including *Health for People* (London: Sampson Low, Marston, Searle, & Rivington, 1886).
30. *Queen*, LXXVI (July–December 1884). See Beetham, *A Magazine of Her Own?*, chapter 7 for *Queen* magazine and its diversity of content.
31. Marland, 'The Diffusion of Useful Knowledge'; Charles Rosenberg (ed.), *Right Living: An Anglo-American Tradition of Self-Help Medicine and Hygiene* (Baltimore, MD and London: Johns Hopkins University Press, 2003).
32. Gordon Stables, *The People's A B C Guide to Health* (London: Hodder and Stoughton, 1887); idem, *Heartstone Talks on Health and Home* (Norwich: Jarrold and Sons, 1904); idem, *The Wife's Guide to Health and Happiness* (London: Jarrold and Sons, 1894); idem, *The Mother's Book of Health and Family Advisor* (London: Jarrold and Sons, 1894); idem, *The Girl's Own Book of Health and Beauty* (London: Jarrold and Sons, 1891); idem, *The Boy's Own Book of Health & Strength* (London: Jarrold and Sons, 1892); idem, *Rota Vitae: The Cyclists Guide to Health and Rational Enjoyment* (London: Iliffe & Son, 1886); idem, *Health upon Wheels; or, Cycling a Means of Maintaining the Health* (London: Iliffe & Son, 1887); idem, *Turkish and other Baths: A Guide to Good Health and Longevity* (London: Dean & Son [1894]).
33. Stables, *The Girl's Own Book of Health and Beauty*; idem, *People's A B C Guide to Health*.
34. George Black, *Household Medicine: A Guide to Good Health, Long Life, and the Proper Treatment of all Diseases and Accidents* (London: Ward, Lock & Co., 1883), pp. 771–4.
35. Malcolm Morris (ed.), *The Book of Health* (London, Paris and New York: Cassell, 1883).
36. Alfred B. Olsen and M. Ellsworth Olsen (eds), *The School of Health: A Guide to Health in the Home* (London: International Tract Society, 1906). See

James C. Whorton, *Crusaders for Fitness: The History of American Health Reformers* (Princeton, NJ: Princeton University Press, 1982) and *idem, Inner Hygiene: Constipation and the Pursuit of Health in Modern Society* (Oxford: Oxford University Press, 2000), for Kellogg's promotion of hygienic practices and vegetarian diet.

37. Galbraith practised medicine in New York and taught at the city's Woman's Medical College and Richards was Medical Director of William Penn High School and an Instructor at the Woman's Medical College of Pennsylvania: Anna M. Galbraith, *Hygiene and Physical Culture for Women* (London: B.F. Stevens, 1895); Richards, *Hygiene for Girls*.

38. Norman Gevitz, ' "But All Those Authors Are Foreigners": American Literary Nationalism and Domestic Health Guides', in Roy Porter (ed.), *The Popularization of Medicine 1650–1850* (London: Routledge, 1992), pp. 232–51; Rosenberg (ed.), *Right Living*.

39. Mary Humphreys, *Personal Hygiene for Girls* (London, New York, Toronto and Melbourne: Cassell, 1913).

40. For women's work in medicine in Britain, see, for example, Mary Ann Elston, ' "Run by Women, (mainly) for Women": Medical Women's Hospitals in Britain, 1866–1948', in Anne Hardy and Lawrence Conrad (eds), *Women and Modern Medicine* (Amsterdam and New York: Rodopi, 2001), pp. 73–107; *idem*, 'Women Doctors in the British Health Services: A Sociological Study of their Careers and Opportunities', unpublished University of Leeds, PhD thesis, 1986.

41. Brian Harrison, 'Women's Health and the Women's Movement in Britain: 1840–1940', in Charles Webster (ed.), *Biology, Medicine and Society 1840–1940* (Cambridge: Cambridge University Press, 1981), pp. 15–71, on pp. 51–2. Harrison gives figures taken from census returns which listed 25 female doctors in England and Wales in 1881 and 495 in 1911.

42. Mathew Thomson, *Psychological Subjects: Identity, Culture, and Health in Twentieth-Century Britain* (Oxford: Oxford University Press, 2006), pp. 48–9. Chesser's books included *Perfect Health for Women and Children* (London: Methuen, 1912); *idem, From Girlhood to Womanhood* (London, New York, Toronto and Melbourne: Cassell, 1913); *idem, Woman, Marriage and Motherhood* (London, New York, Toronto and Melbourne: Cassell, 1913); *idem, Physiology and Hygiene for Girls' Schools and Colleges* (London: G. Bell and Sons, 1914); *idem, Vitality: A Book on the Health of Women and Children* (London: Methuen, 1935).

43. See Mitchell, *The New Girl*, chapter 1 for the drift away from mother-daughter literature. For advice to mothers and daughters, see Deborah Gorham, *The Victorian Girl and the Feminine Ideal* (London and Canberra: Croom Helm, 1982). Publications dedicated to motherhood, pregnancy and the management of the lying-in room, female disorders, infant and child care and home nursing also endured. Dr Thomas Bull's *Hints to Mothers* remained popular throughout the late nineteenth century, appearing in its 27th edition in 1879, as did magazines such as *The Mother's Friend*, published from 1845 to 1895. For a selection of this literature, see Ruth Robbins (ed.), *Medical Advice for Women, 1830–1915* (Abingdon: Routledge, 2008–). For health advice for mothers, see Lisa Petermann, 'From a Cough to a Coffin: The Child's Medical Experience in Britain and France, 1762–1884',

unpublished University of Warwick PhD thesis, 2007; Patricia Branca, *Silent Sisterhood: Middle-Class Women in the Victorian Home* (London: Croom Helm, 1975), chapters 5 and 6; Christina Hardyment, *Dream Babies: Child Care from Locke to Spock* (London: Jonathan Cape, 1983), especially chapter 3 which explores the 'renaissance' of motherhood between 1870 and 1920; Lyubov G. Gurjeva, 'Child Health, Commerce and Family Values: The Domestic Production of the Middle Class in Late-Nineteenth and Early-Twentieth Century Britain', in Marijke Gijswijt-Hofstra and Hilary Marland (eds), *Cultures of Child Health in Britain and the Netherlands in the Twentieth Century* (Amsterdam and New York: Rodopi, 2003), pp. 103–25; and for the US Rima D. Apple, *Mothers and Medicine: A Social History of Infant Feeding 1890–1950* (Madison, WI: University of Wisconsin Press, 1987).

44. Pye Henry Chavasse, *Advice to a Mother on the Management of Her Children and on the Treatment on the Moment of Some of Their More Pressing Illnesses and Accidents*, 14th edn (London: J. & A. Churchill, 1889), preface. The book reached its fourteenth edition in 1889, by which time Chavasse had died and the editorship had been taken over by a series of 'guest' editors. The book had sold 120,000 copies and, according to its publishers, was widely read everywhere where the English language was spoken. It was still being published in the 1930s with different editors, revisions and slight variations in title, adapting to changing social and cultural expectations of health, youth and parenting.

45. See Julie-Marie Strange, 'The Assault on Ignorance: Teaching Menstrual Etiquette in England, *c*.1920s-1960s', *Social History of Medicine*, 14 (2001), 247–65.

46. Joan Jacobs Brumberg, ' "Something Happens to Girls": Menarche and the Emergence of the Modern Hygienic Imperative', in Judy Walzer Leavitt (ed.), *Women and Health in America*, 2nd edn (Madison, WI: University of Wisconsin Press, 1999), pp. 150–71, on p. 153. See also Elaine Showalter and English Showalter, 'Victorian Women and Menstruation', in Martha Vicinus (ed.), *Suffer and Be Still: Women in the Victorian Age* (London: Methuen, 1980; first published Bloomington, IN: Indiana University Press, 1972), pp. 38–44.

47. E. J. Tilt, *Elements of Health, and Principles of Female Hygiene* (London: Henry G. Bohn, 1852), p. 179. See also Ali Al-Khalidi, 'Emergent Technologies in Menstrual Paraphernalia in Mid-Nineteenth-Century Britain', *Journal of Design History*, 14 (2001), 257–73, especially p. 258.

48. Lionel Weatherly, *The Young Wife's Own Book: A Manual of Personal and Family Hygiene, Containing Everything That the Young Wife and Mother Ought to Know Concerning Her Own Health and that of Her Children at the Most Important Periods of Life* (London: Griffith and Farran, 1882; published simultaneously in New York: E.P. Dutton & Co.); George Black, *The Young Wife's Advice Book: A Guide for Mothers on Health and Self-Management*, 6th edn (London: Ward, Lock and Co., 1888); Mary Lynn Stewart, *For Health and Beauty: Physical Culture for Frenchwomen, 1880s-1930s* (Baltimore, MD and London: Johns Hopkins University Press, 2001), chapter 4, quote on p. 78. See, for broader notions of women and pollution, Alison Bashford, *Purity and Pollution: Gender, Embodiment and Victorian Medicine* (Houndmills: Macmillan, 1998), p. 37.

49. J.H. Walsh, *Domestic Medicine and Surgery: With a Glossary of the Terms Used Therein* (London: Frederick Warne, 1866), pp. 167, 468. Walsh, a general practitioner and between 1849 and 1852 editor of the *Provincial Medical and Surgical Journal*, published widely, often under the pseudonym 'Stonehenge', on domestic economy, cookery, field sports, dog breeding and horses. G.C. Boase, *rev.* Julian Lock, 'Walsh, John Henry [*pseud*. Stonehenge] (1810–1888)', *Oxford Dictionary of National Biography* (hereafter *DNB*): [http://0www.oxforddnb.com.pugwash.lib.warwick.ac.uk/view/article/28614, accessed 8 February 2012].

50. George Black, *The Young Wife's Advice Book: A Complete Guide for Mothers on Health, Self-Management, and the Care of the Baby* (London, Melbourne and Toronto: Ward, Lock and Co., 1910), pp. 288–9.

51. Weatherly, *The Young Wife's Own Book*, p. 29.

52. Ibid., p. 28.

53. Ibid., p. 30.

54. Ibid., p. 28.

55. Black, *The Young Wife's Advice Book* (6th edn, 1888), p. 6.

56. Ibid., p. 5.

57. A. Reeves Jackson, 'Diseases Peculiar to Women', in Frederick A. Castle (ed.), *Wood's Household Practice of Medicine Hygiene and Surgery: A Practical Treatise for the Use of Families, Travellers, Seamen, Miners and Others*, vol. 2 (London: Sampson Low, Marsten, Searle, & Rivington, 1881), pp. 538–9.

58. Ibid., pp. 539, 542, 543.

59. Stables, *The Wife's Guide to Health and Happiness*, p. 119.

60. Ibid., pp. 121–3.

61. Helen Corke, *In Our Infancy: An Autobiography, Part I: 1882–1912* (Cambridge: Cambridge University Press, 1975), p. 93.

62. Ibid.

63. Edward Bruce Kirk, *A Talk with Girls about Themselves* (London: Simpkin, Marshall, Hamilton, Kent & Cov., 1895), p. 47a.

64. Chesser, *Physiology and Hygiene*, p. 86.

65. Anon., 'Chlorosis: A Disease of Early Womanhood', *Good Health*, VIII: 192 (4 July 1896), 418–19.

66. Pye Henry Chavasse, *Advice to a Mother: On the Management of Her Children, and on the Treatment on the Moment of Some of their More Pressing Illnesses and Accidents*, 9th edn (London: Cassell, 1911), p. 298. The editions of Chavasse's volume are not sequential in terms of date.

67. Curwen and Herbert, *Simple Health Rules*, p. 10.

68. Margaret Hallam, *Health and Beauty for Women and Girls: A Course of Physical Culture* (London: C. Arthur Pearson, 1921), with a preface by eugenicist physician Mary Scharlieb, pp. 110–17, quotes on pp. 111, 112.

69. Emma E. Walker, *Beauty through Hygiene: Common-Sense Ways to Health for Girls* (London: Hutchinson & Co., 1905), pp. 185–90, quote on p. 190.

70. Flint, *The Woman Reader*, p. 92; Marianne Farningham, *Girlhood* (London: James Clarke, 1895).

71. Stables, *The Girl's Own Book of Health and Beauty*, p. 131.

72. Gorham, *The Victorian Girl*, p. 93.

73. 'Medicus', 'Common-Sense Advice for Working Girls', *GOP*, VI: 267 (7 February 1885), 295–6; Curwen and Herbert, *Simple Health Rules*, preface, p. 1.

74. Anon., 'Talks with Girls: Some Healthy Habits', *Good Health*, VI: 155 (21 September 1895), 786–7.
75. Stables, *The Girl's Own Book of Health and Beauty*, preface.
76. Walker, *Beauty through Hygiene*, p. 18. The importance of correct breathing was emphasised by many authors and became an important aspect of the physical culture movement. Margaret Hallam, author of several guides to health and beauty, explained that 'the first step in Physical Culture is to learn how to breathe properly': *Health and Beauty for Women and Girls*, p. 24.
77. Stables, *The Girl's Own Book of Health and Beauty*, chapter 1, quotes on pp. 17, 19.
78. Walker, *Beauty through Hygiene*, p. 10.
79. Chesser, *From Girlhood to Womanhood*, pp. 45–6.
80. Agnes L. Stenhouse and E. Stenhouse, *A Health Reader for Girls* (London: Macmillan, 1918), p. 45 (their emphasis).
81. Clement Dukes, 'The Hygiene of Youth', in T.C. Allbutt and H.D. Rolleston (eds), *A System of Medicine*, vol. 1 (London and New York: Macmillan, 1905), pp. 161–81, quote on p. 179.
82. Clement Dukes, 'Health at School', in Morris (ed.), *The Book of Health*, pp. 677–725, quote on p. 724.
83. Chesser, *From Girlhood to Womanhood*, pp. 77–8.
84. Thomson, *Psychological Subjects*, pp. 47–51.
85. Ibid., pp. 140–9.
86. Chavasse, *Advice to a Mother* (14th edn, 1889), pp. 322–7, quote on p. 326.
87. Ibid., p. 321.
88. Humphreys, *Personal Hygiene for Girls*, pp. 14–15 (her emphasis).
89. Chesser, *Physiology and Hygiene*, pp. 86, 84.
90. Amy B. Barnard, *The Girl's Encyclopaedia* (London: Pilgrim, 1909), p. 3.
91. Dr Andrew Wilson, 'Plain Talks about Common Ailments and their Cures', *Health* (11 January 1895), 226–7, quote on p. 226.
92. A.T. Schofield, 'Nervousness and Hysteria', *The Leisure Hour* (June 1889), 412–15, quote on p. 415. Schofield's account closely mirrors that of F.C. Skey in *Hysteria: Six Lectures Delivered to the Students of St. Bartholomew's Hospital, 1866* (London: Longmans, Green, Reader & Dyer, 1867), p. 55.
93. Stenhouse and Stenhouse, *A Health Reader for Girls*, pp. 180, 182.
94. Curwen and Herbert, *Simple Health Rules*, pp. 7–10, quote on p. 10.
95. Walker, *Beauty Through Hygiene*, pp. 298–305, quotes on pp. 304, 305.
96. Hallam, *Health and Beauty for Women and Girls*, p. 117.
97. Anna Bonus Kingsford, *Health, Beauty and the Toilet: Letters to Ladies from a Lady Doctor* (London and New York: Frederick Warne, 1886).
98. Ibid., preface.
99. Ibid.; Lori Williamson, 'Kingsford [née Bonus], Anna [Annie] (1846–1888)', *DNB*: [http://0-www.oxforddnb.com.pugwash.lib.warwick.ac.uk/view/article/15615, accessed 7 February 2012].
100. A Specialist [Anon.], *Beauty and Hygiene for Women and Girls* (London: Swan Sonnenschein & Co., 1893), quote on p. 1. Defining Gender, 1450–1910, Matthew Adams Digital, Bodleian Library 1893, 17502 e.8.
101. Kingsford, *Health, Beauty and the Toilet*, p. 121.
102. Ibid., pp. 121–2.
103. Stables, *The Girl's Own Book of Health and Beauty*, preface.

104. Anon., 'Beauty Hints for Women: Talks with Girls', *Good Health*, VII: 161 (2 November 1895), 141; *idem*, 'Hints for Women Concerning Beauty', *Good Health*, VI: 141 (15 June 1895), 337.
105. Barnard, *The Girl's Encyclopaedia*, p. 3.
106. Stables, *The Girl's Own Book of Health and Beauty*, pp. 36–49, 147–8.
107. 'Medicus', 'Health and Beauty for the Hair', *GOP*, I: 17 (24 April 1880), 259–60, quote on p. 260.
108. Curwen and Herbert, *Simple Health Rules*, p. 11.
109. Stenhouse and Stenhouse, *A Health Reader for Girls*, p. 110.
110. Chesser, *Physiology and Hygiene*, p. 79.
111. Richards, *Hygiene for Girls*, pp. 7–8.
112. Ibid.
113. Humphreys, *Personal Hygiene for Girls*, p. 50.
114. Hallam, *Health and Beauty for Women and Girls*, pp. 75–83, quotes on pp. 76, 77. See also *idem*, *Dear Daughter of Eve: A Complete Book of Health and Beauty* (London: W. Collins, 1924).
115. Humphreys, *Personal Hygiene for Girls*, pp. 81, 86.
116. Chesser, *Physiology and Hygiene*.
117. Annie Burns Smith, *Talks with Girls upon Personal Hygiene* (London: Pitman, 1912), with a Preface by John Robertson, Medical Officer of Health, City of Birmingham, and an Introduction by Mary D. Sturge, Physician to the Birmingham and Midland Hospital for Women, p. 11. For the impact of germs on American women in the home, see Nancy Tomes, *The Gospel of Germs: Men, Women and the Microbe in American Life* (Cambridge, MA and London: Harvard University Press, 1998); Barbara Ehrenreich and Deirdre English, *For Her Own Good: 150 Years of the Experts' Advice to Women* (New York: Anchor/Doubleday, 1978), chapter 5.
118. Smith, *Talks with Girls Upon Personal Hygiene*.
119. See for Macfadden, Whorton, *Crusaders for Fitness*, pp. 296–303 and Mark Adams, *Mr. America* (New York: Harper, 2009) and for Sandow, Ueyama, *Health in the Marketplace*, pp. 102–6.
120. Nanette M. Pratt, 'Health Talk', *Woman's Beauty and Health* (Special Edition for England), I: 2 (May 1902), 47–9, quote on p. 47. Defining Gender, 1450–1910, 1902 per 17502 d.4).
121. Zweiniger-Bargielowska, *Managing the Body*, p. 142.
122. Cheryl Buckley and Hilary Fawcett, *Fashioning the Feminine: Representation and Women's Fashion from the Fin de Siècle to the Present* (London: I.B. Taurus, 2002), p. 35.
123. For example, Stenhouse and Stenhouse, *A Health Reader for Girls*, p. 118.
124. Anon., 'Athletic Girls and Good Looks', *The Domestic Magazine and Journal for the Household*, V: 121 (26 January 1895), 263. The first best-selling diet guide in the US was William Banting's *Letter on Corpulence* (1863), which went through twelve editions by 1900. See, for diet regimes, Katharina Vester, 'Regime Change: Gender, Class, and the Invention of Dieting in Post-Bellum America', *Journal of Social History*, 44 (2010), 39–70 and Hillel Schwartz, *Never Satisfied: A Cultural History of Diets, Fantasies and Fat* (New York: Free Press, 1986).
125. For the renaissance of auto-intoxication theory in the late nineteenth century and early twentieth century, see Ina Zweiniger-Bargileowska, 'Raising a

Nation of "Good Animals": The New Health Society and Health Education Campaigns in Interwar Britain', *Social History of Medicine*, 20 (2007), 73–89. This explores the work of Sir William Arbuthnot Lane who saw the bowels as central to health and constipation as a root cause of the ills of civilisation. See also Whorton, *Inner Hygiene: Constipation and the Pursuit of Health in Modern Society*.

126. Kingsford, *Health, Beauty and the Toilet*, p. 12.
127. Vester, 'Regime Change', p. 48.
128. Chesser, *Physiology and Hygiene*, p. 83.
129. Kingsford, *Health, Beauty and the Toilet*, p. 10.
130. Walker, *Beauty through Hygiene*, chapter VI.
131. Stables, *The Girl's Own Book of Health and Beauty*, pp. 174–82.
132. Gorham, *The Victorian Girl*, p. 93.
133. See Buckley and Fawcett, *Fashioning the Feminine*.
134. Walker, *Beauty through Hygiene*, p. 273.
135. Olsen and Olsen, *The School of Health*, pp. 109–10. There is a rich literature on dress reform and attitudes to corsets, including Valerie Steel, *The Corset: A Cultural History* (New Haven, CT and London: Yale University Press, 2001) and Leigh Summers, *Bound to Please: A History of the Victorian Corset* (Oxford: Berg, 2001). See also Stella Mary Newton, *Health, Art and Reason: Dress Reformers of the 19th Century* (London: John Murray, 1974).
136. Frederick Treves, 'The Influence of Dress on Health', in Morris (ed.), *Book of Health*, pp. 461–517, especially pp. 496–506 for tight lacing.
137. Smith, *Talks with Girls upon Personal Hygiene*, pp. 109–10, quote on p. 110.
138. Ibid., pp. 110–11.
139. Howard Spicer (ed.), *Sports for Girls*, with an introduction by Mrs Ada S. Ballin (London: Andrew Melrose, 1900), p. 8.
140. Chesser, *Physiology and Hygiene*, p. 149.
141. Humphreys, *Personal Hygiene for Girls*, pp. 89, 94–5 (her emphasis).
142. Stenhouse and Stenhouse, *A Health Reader for Girls*, pp. 118–24.
143. Chesser, *Physiology and Hygiene*, p. 146.
144. See for the GOP, Wendy Forrester, *Great-Grandmama's Weekly: A Celebration of The Girl's Own Paper 1880–1901* (Guildford and London: Lutterworth Press, 1980); Doughty (ed.), *Selections from the Girl's Own Paper*; Hilary Skelding, 'Every Girl's Best Friend? The *Girl's Own Paper* and its Readers', in Emma Liggins and Daniel Duffy (eds), *Feminist Readings of Victorian Popular Texts: Divergent Femininities* (Aldershot: Ashgate, 2001), pp. 35–52; Drotner, *English Children and their Magazines*, especially chapter 10; and for the relationship between girlhood and Empire, Michelle Smith, *Empire in British Girls' Literature and Culture: Imperial Girls, 1880–1915* (Houndmills: Palgrave Macmillan, 2011), chapter 1.
145. Drotner, *English Children and Their Magazines*, p. 116; Beetham, *A Magazine of Her Own?*, chapter 8.
146. 'Answers to Correspondents', GOP, II: 40 (2 October 1880), 15.
147. Forrester, *Great-Grandmama's Weekly*, p. 14.
148. Doughty (ed.), *Selections from the Girl's Own Paper*, p. 8; Kimberley Reynolds, *Girls Only? Gender and Popular Children's Fiction in Britain, 1880–1910* (Hemel Hempstead: Harvester Wheatsheaf, 1990), p. 150.
149. Tinkler, *Constructing Girlhood*, pp. 49–50.

150. For example, A.T. Schofield, 'A New Career for Ladies [public health work]', *GOP*, XII: 612 (19 September 1891), 809–10; S.F.A. Caulfeild, 'The Progress of Women's Work', *GOP*, XV: 723 (4 November 1893), 77–8; Anon., 'The Rise of a New Profession [physical education] for Girls', *GOP*, XV: 733 (13 January 1894), 249.

151. Tinkler, *Constructing Girlhood*, pp. 48–9. See Skelding, 'Every Girl's Best Friend?' for examples of this diversity of material and approaches.

152. Drotner, *English Children and their Magazines*, p. 116.

153. F.H [Frank Hird], 'Women's Work', p. 51.

154. A.T. Schofield, 'On the Perfecting of the Modern Girl', *GOP*, XVI: 805 (1 June 1895), 662.

155. S.F.A. Caulfeild, 'Some Types of Girlhood; or, Our Juvenile Spinsters', *GOP*, XII: 562 (4 October 1890), 4–5.

156. Lucy H. Yates, 'The Girls of To-day', *GOP*, XIV: 764 (18 August 1894), 724–5.

157. For replies to correspondents, see Roy Hindle, *Oh, No Dear! Advice to Girls a Century Ago* (Newton Abbot: David & Charles, 1982). The 500 replies were selected from the years 1880–1882 out of a total of 10,000, and many dealt with health matters.

158. 'Varieties', *GOP*, VI: 296 (29 August 1885), 759.

159. 'Varieties', *GOP*, VII: 331 (1 May 1886), 494.

160. 'Answers to Correspondents', *GOP*, IX: 422 (28 January 1888), 288.

161. 'Answers to Correspondents', *GOP*, XX: 980 (8 October 1898), 30.

162. 'Answers to Correspondents', *GOP*, XXI: 1041 (9 December 1899), 159.

163. Stables, *The Girl's Own Book of Health and Beauty*, preface.

164. Anon., 'How to be Healthy, Happy, and Beautiful', *GOP*, II: 44 (30 October 1880), 66–7, quote on p. 67.

165. 'Medicus', 'The Weather and Health', *GOP*, VIII: 354 (9 October 1886), 23–4, quote on p. 23.

166. 'Medicus', 'Can Girls Increase Their Strength?', *GOP*, XV: 752 (26 May 1894), 534.

167. 'Medicus', 'A New Year's Health Sermon', *GOP*, XVII: 836 (4 January 1896), 211.

168. 'Medicus', 'Health, Strength, and Beauty', *GOP*, XI: 557 (30 August 1890), 758–9, quote on p. 758.

169. 'Medicus' (Dr. Gordon-Stables, R.N.), 'Physical Culture for Girls', *GOP*, XXI: 1037 (11 November 1899), 86–7, quote on p. 86.

170. 'Medicus', 'Health, Strength, and Beauty', p. 758.

171. 'Medicus', 'Health All the Year Round', *GOP*, VI: 259 (13 December 1884), 166–7, quote on p. 166.

172. Drotner, *English Children and their Magazines*, pp. 152–3.

173. 'Medicus', 'Common-Sense Advice for Working Girls', pp. 295–6.

174. 'Medicus', 'A Plain Talk with Sensible Girls', *GOP*, XVII: 860 (20 June 1896), 597.

175. 'Medicus', 'Health in the Kitchen-Garden', *GOP*, VIII: 375 (5 March 1887), 356–8, quote on p. 357.

176. Mrs Wallace Arnold, 'The Physical Education of Girls', *GOP*, V: 229 (17 May 1884), 516–17.

177. 'The Editor of "Physical Culture"', 'The Physical Training of Girls', *GOP*, XXI: 1058 (7 April 1900), 422–3, 'Medicus', 'Physical Culture for Girls'.

178. 'The New Doctor', 'Exercise: In Moderation', *GOP*, XX: 999 (18 February 1899), 326–7, quote on p. 326.
179. Arnold, 'The Physical Education of Girls', 516–17.
180. Ibid., p. 516.
181. Anon., 'The Girl of To-day', *GOP*, XXXIX: 2 (1916–17, no issue date), 85.
182. Annette Kellerman, *Physical Beauty and How to Keep It* (London: Heinemann, 1918). Cited in Jill Julius Matthews, 'Building the Body Beautiful', *Australian Feminist Studies*, 5 (1987), 17–34, on p. 26. See, for example, Bernarr MacFadden and Marion Malcolm, *Health – Beauty – Sexuality: From Girlhood to Womanhood* (New York: Physical Culture Publishing Company, 1904).
183. Mary Scharlieb, *Venereal Diseases in Children and Adolescents: Their Recognition and Prevention (Notes of Three Lectures addressed to Schoolmistresses at the Royal Society of Medicine, London, September, 1916)* (London: National Council for Combating Venereal Diseases, 1916), p. 9.
184. Curwen and Herbert, *Simple Health Rules*, p. 15.
185. Chesser, *Physiology and Hygiene*, p. 83.
186. Stearns and Stearns, *Anger*, p. 36.
187. A.B. Barnard, *The Girl's Book About Herself* (London: Cassell, 1912), p. 11.
188. Ibid., p. 18.

3 Health, Exercise and the Emergence of the Modern Girl

1. Arabella Kenealy, 'Woman as Athlete', *The Nineteenth Century* (April 1899), 635–45, quotes on pp. 635, 641. For Kenealy, see Angelique Richardson, 'Kenealy, Arabella Madonna (1859–1938)', *Oxford Dictionary of National Biography* (hereafter *DNB*): [http://0-www.oxforddnb.com. pugwash.lib.warwick.ac.uk/view/article/50057, accessed 31 October 2011].
2. Kenealy, 'Woman as Athlete', p. 639.
3. Ina Zweiniger-Bargielowska cites a comparable example of the story of Miggins' wife Angela, who fell prone to the 'disease' of physical culture, turning the household upside down and emasculating Miggins: *Health and Strength* (September 1904), 110–14. Cited in Ina Zweiniger-Bargielowska, *Managing the Body: Beauty, Health, and Fitness in Britain, 1880–1939* (Oxford: Oxford University Press, 2010), pp. 105–6.
4. Kenealy, 'Woman as Athlete', p. 643.
5. As referred to in Chapter 1: W.S. Playfair, 'The Nervous System in Relation to Gynaecology', in Thomas Clifford Allbutt and W.S. Playfair (eds), *A System of Gynaecology by Many Writers* (London: Macmillan, 1896), pp. 220–31, on pp. 221.
6. Anon., 'The Month', *Practitioner*, LV (August 1895), 111.
7. See Patricia Marks, *Bicycles, Bangs, and Bloomers: The New Woman in the Popular Press* (Lexington: The University Press of Kentucky, 1990), Chapter 6.
8. L. Ormiston Chant, 'Woman as an Athlete: A Reply to Dr. Arabella Kenealy', *The Nineteenth Century* (May 1899), 745–54, especially p. 748.
9. Arabella Kenealy, 'Woman as an Athlete: A Rejoinder', *The Nineteenth Century* (June 1899), 915–29, quotes on pp. 916, 920. See, for Kenealy's views

on the extinction of sex in adolescence, her *Feminism and Sex-Extinction* (London: T. Fisher Unwin [1920]).

10. Kenealy, 'Woman as an Athlete: A Rejoinder', pp. 924, 926, 928.

11. Kenealy, *Feminism and Sex-Extinction*, p. 118.

12. A. Shadwell, 'The Hidden Dangers of Cycling', *The National Review*, 28: 168 (February 1897), 787–96, especially p. 795; James C. Whorton, 'The Hygiene of the Wheel: An Episode in Victorian Sanitary Science', *Bulletin of the History of Medicine*, 52 (1978), 61–88, quote on pp. 84–5.

13. Bruce Haley, *The Healthy Body and Victorian Culture* (Cambridge, MA and London: Harvard University Press, 1978). See also J.A. Mangan, *Athleticism in the Victorian and Edwardian Public School* (Cambridge: Cambridge University Press, 1981), and for women's take up of sport, Kathleen E. McCrone, *Sport and the Physical Emancipation of English Women 1870–1914* (London: Routledge, 1988); Sheila Fletcher, *Women First: The Female Tradition in English Physical Education 1800–1980* (London: Athlone, 1984); Jennifer Hargreaves, *Sporting Females: Critical Issues in the History and Sociology of Women's Sports* (London: Routledge, 1994); Patricia A. Vertinsky, *The Eternally Wounded Woman: Women, Doctors, and Exercise in the Late Nineteenth Century* (Urbana and Chicago, IL: University of Illinois Press, 1994; first published Manchester University Press, 1989) and *idem*, 'Body Shapes: The Role of the Medical Establishment in Informing Female Exercise and Physical Education in Nineteenth-Century North America', in J.A. Mangan and Roberta J. Parke (eds), *From 'Fair Sex' to Feminism: Sport and the Socialization of Women in the Industrial and Post-Industrial Eras* (London: Frank Cass, 1987), pp. 256–81.

14. The cycle became a popular mode of transport for doctors during the 1880s. See P.W.J. Bartrip, *Mirror of Medicine: History of the BMJ* (Oxford: British Medical Journal and Clarendon Press, 1990), pp. 148–51, for coverage of cycling in the *BMJ*.

15. Anon., 'Modern Types: No. XXII – The Manly Maiden', *Punch*, 99 (6 December 1890), 265; Anon., 'Modern Mannish Maidens', *Blackwood's Magazine*, 252–64.

16. Jonathan May, *Madame Bergman-Österberg: Pioneer of Physical Education and Games for Girls and Women* (London: George G. Harrap, 1969), p. 39; Takahiro Ueyama, *Health in the Marketplace: Professionalism, Therapeutic Desires, and Medical Commodification in Late-Victorian London* (Palo Alto, CA: Society for the Promotion of Science and Scholarship, 2010), pp. 100–6.

17. Bessie Rayner Parkes, *Remarks on the Education of Girls* (London: John Chapman, 1854), p. 8. Cited in McCrone, *Sport and the Physical Emancipation of English Women*, p. 15.

18. Anon., 'Swimming for Women', *The London Review*, 8 (25 August 1860), 182; Harriet Martineau, 'How to Learn to Swim', *Once a Week*, 1: 16 (15 October 1859), pp. 327–8. See also Sally Mitchell, *The New Girl: Girls' Culture in England 1880–1915* (New York: Columbia University Press, 1995), pp. 109–10. Catherine Horwood, ' "Girls Who Arouse Dangerous Passions": Women and Bathing, 1900–39', *Women's History Review*, 9 (2000), 653–73 describes women's struggle to gain access to swimming pools and beaches in Britain between 1900 and 1939.

19. Horwood, ' "Girls Who Arouse Dangerous Passions" ', p. 654; McCrone, *Sport and the Physical Emancipation of English Women*; Fletcher, *Women First*.
20. See Mitchell, *The New Girl*, especially Chapter 5.
21. Anon., 'Girls Physical Training', *Health* (8 November 1895), 85.
22. E.M. Symonds, 'Gymnastics for Girls', *The Girl's Realm Annual* (November 1898–October 1899) (London: Hutchinson, 1899), pp. 154–7, on p. 154.
23. Anon., 'The Woman of the Future', *Good Health*, VI: 141 (15 June 1895), 325.
24. Pye Henry Chavasse, *Advice to a Mother on the Management of Her Children and on the Treatment on the Moment of Some of their More Pressing Illnesses and Accidents*, 14th edn (London: J. & A. Churchill, 1889), p. 286.
25. Mary Whitley, *Every Girl's Book of Sport, Occupation and Pastime* (London: George Routledge and Sons, 1897).
26. Emma E. Walker, *Beauty through Hygiene: Common-Sense Ways to Health for Girls* (London: Hutchinson, 1905), quote on p. 61.
27. Anon., 'Exercise for Women', *Good Health*, IV: 100 (1 September 1894), 290; James Cantlie, 'The Influence of Exercise on Health', in Malcolm Morris (ed.), *The Book of Health* (London, Paris and New York: Cassell, 1883), pp. 381–461, on p. 397.
28. McCrone, *Sport and the Physical Emancipation of English Women*, p. 193.
29. J. Hamilton Fletcher, 'Feminine Athletics', *Good Words*, 20 (January 1879), 533–6, quote on p. 534. See also Tracy J.R. Collins, 'Physical Fitness, Sports, Athletics and the Rise of the New Woman', unpublished University of Purdue PhD thesis, 2007, pp. 43–5.
30. E.D. Bourne (ed.), *Girls' Games: A Recreation Handbook for Teachers and Scholars* (London: Griffith, Farran, Okeden & Welsh, 1887), pp. 1–2.
31. Alfred T. Schofield, 'The Modern Development of Athletics', *The Leisure Hour* (October 1891), 817–20, quotes on p. 818.
32. Mary Taylor Bissell, 'Athletics for City Girls', *Health* (15 February, 1895), 310–11 (22 February, 1895), 327–8, first published *Popular Science Monthly*, 46 (December 1894), 146–53. It was common practice to reprint articles that had appeared in overseas, notably US, journals.
33. Ibid., p. 327.
34. Lawrence H. Prince, 'Physical Exercise for Women: From a Medical Point of View', *Good Health*, VI: 138 (25 May 1895), 230–1; VI: 141 (15 June 1895), 326–7; VI: 142 (22 June 1895), 356–7, quote on p. 327.
35. Schofield, 'The Modern Development of Athletics', p. 819.
36. Ibid., p. 818.
37. Ernest W. Lowe, 'Women and Exercise', *Chambers's Journal*, 3: 108 (23 December 1899), 49–52, quotes on pp. 50, 49.
38. Ibid., p. 50.
39. She went on to set up Dartford Physical Education College in 1895: Sheila Fletcher, 'Österberg, Martina Sofia Helena Bergman (1849–1915)', *DNB*: [http://0-www.oxforddnb.com.pugwash.lib.warwick.ac.uk/view/article/47656, accessed 9 March 2012].
40. Anon., 'Interview with Madame Bergman-Österberg', *Woman's Herald* (20 June 1891). Cited in Fletcher, *Women First*, pp. 23–4. For Bergman Österberg's influence on physical education, see *idem*; May, *Madame Bergman-Österberg* and Ida M. Webb, 'Women's Physical Education in Great Britain,

1800–1966, with Special Reference to Teacher Training', unpublished University of Leicester MEd thesis, 1967.

41. Bim Andrews, 'Making Do'. Cited in John Burnett (ed.), *Destiny Obscure: Autobiographies of Childhood, Education and Family from the 1820s to the 1920s* (London and New York: Routledge, 1982), pp. 120, 123.

42. Anon. ['Medicus', Gordon Stables], 'How to be Healthy, Happy and Beautiful', *Girl's Own Paper (GOP)*, II: 44 (30 October 1880), 66–7, on p. 67.

43. Richard Harmond, 'Progress and Flight: An Interpretation of the American Cycle Craze of the 1890s', *Journal of Social History*, 5 (Winter, 1971–72), 235–57, on p. 244.

44. Madame Marie, 'Housework as a Beautifier', *London Journal*, 7: 164 (12 June 1909), 165.

45. Anne Mahon, 'Housework and Health: Daily Toil as Physical Drill', *GOP*, XXXVII: 8 (1915–16, no individual dates for issues), 465–6, quote on p. 465.

46. Cantlie, 'The Influence of Exercise on Health' offered a comprehensive survey of the dangers of overexertion and the importance of moderation and training in a range of sports for boys and girls.

47. Lowe, 'Women and Exercise', p. 50.

48. Walker, *Beauty through Hygiene*, preface, quote on p. 11.

49. Whitley, *Every Girl's Book of Sport, Occupation and Pastime*, pp. 1–2.

50. Anne M. Sebba, 'Ballin, Ada Sarah (1862–1906)', *DNB*: [http://0-www.oxforddnb.com.pugwash.lib.warwick.ac.uk/view/article/55732, accessed 20 October 2011]. Ballin also founded *Baby: The Mothers' Magazine* in 1887 and wrote a range of sixpenny pamphlets in her Mothers' Guide series.

51. Howard Spicer (ed.), *Sports for Girls*, with an introduction by Ada S. Ballin (London: Andrew Melrose, 1900), pp. 15, 7.

52. Sir Benjamin Ward Richardson, 'On Recreation for Girls', *GOP*, XV: 753 (2 June 1894), 545–7, p. 546; Patrick Wallis, 'Richardson, Sir Benjamin Ward (1828–1896)', *DNB*: http://0-www.oxforddnb.com.pugwash.lib.warwick.ac.uk/view/article/23544, accessed 27 October 2011]; Benjamin Ward Richardson, *Health and Life* (London: Daldy, Isbester, 1878).

53. Richardson, 'On Recreation for Girls', pp. 546–7.

54. Charles Peters (ed.), *The Girl's Own Outdoor Book Containing Practical Help to Girls on Matters relating to Outdoor Occupation and Recreation* (London: The Religious Tract Society, 1889).

55. Ibid., p. 24.

56. 'Medicus', 'Can Girls Increase their Strength?', *GOP*, XV: 752 (26 May 1894), 533–4, quote on p. 534.

57. Amy B. Barnard, *The Girl's Encyclopaedia* (London: Pilgrim, 1909), p. 16.

58. Ibid., p. 15.

59. Ueyama, *Health in the Marketplace*, pp. 102–6.

60. For example, 'Medicus', 'Physical Culture for Girls', *GOP*, XXI: 1037 (11 November 1899), 86–7; 'The Editor of "Physical Culture"', The Physical Training of Girls', *GOP*, XXI: 1058 (7 April 1900), 422–3.

61. *The Girl's Realm Annual* (November 1898–October 1899) and (November 1899–October 1990) (London: S.H. Bousfield, 1900).

62. Symonds, 'Gymnastics for Girls', pp. 154–7, quote on p. 154.

63. Ibid., p. 157.

64. Fletcher, 'Feminine Athletics', p. 536.

65. E.M. Symonds, 'Physical Culture for Girls', *The Girl's Realm Annual* (November 1898–October 1899), pp. 59–63.
66. Agnes L. Stenhouse and E. Stenhouse, *A Health Reader for Girls* (London: Macmillan, 1918), pp. 34–5.
67. Mary Humphreys, *Personal Hygiene for Girls* (London, New York, Toronto and Melbourne: Cassell, 1913), p. 123.
68. Barnard, *The Girl's Encyclopaedia*, p. 14.
69. Christine Terhune Herrick, 'Women in Athletics', *The Idler* (October 1902), 677–80, quote on p. 680.
70. Modern Records Centre, University of Warwick (hereafter MRC), National Cycle Archive holds a vast array of records on cycle associations and runs of cycle periodicals, which have been drawn on extensively here.
71. See E. Wilson and E. Taylor, *Through the Looking Glass: A History of Dress from 1860 to the Present Day* (London: BBC Books, 1989); Elizabeth Wilson, *Adorned in Dreams: Fashion and Modernity* (London: Virago, 1985); Jennifer Craik, *The Faces of Fashion: Cultural Studies in Fashion* (London: Routledge, 1994); McCrone, *Sport and the Physical Emancipation of Women*, Chapter 8.
72. David Rubinstein, 'Cycling in the 1890s', *Victorian Studies*, 21 (Autumn, 1977), 47–71, on pp. 61–2. See also Marilyn Bonnell, 'The Power of the Pedal: The Bicycle and the Turn-of-the-Century Woman', *Nineteenth-Century Contexts: An Interdisciplinary Journal*, 14 (1990), 215–39.
73. MBS, 'Cycling as an Important Factor in the Higher Education of Women', *Queen*, XCVIII (July–December 1895) (27 July 1895), 193.
74. Mrs A. Tweedie, 'Bicycling versus "Crise de Nerves"', *Queen*, XCXI (July–December 1898) (21 July 1898), 38.
75. Eliza Lynn Linton, 'The Cycling Craze for Ladies', *Lady's Realm* (December 1896), 173–7, quotes on pp. 173, 177.
76. Rubinstein, 'Cycling in the 1890s', p. 51; Peter N. Stearns, 'Stages of Consumerism: Recent Work on the Issues of Periodization', *Journal of Modern History*, 69 (March 1997), 102–17, on p. 109.
77. Rubinstein, 'Cycling in the 1890s', pp. 49–50.
78. Anon., 'British Medical Journal Reports: A Report on Cycling in Health and Disease', *British Medical Journal (BMJ)*, 1 (9 May 1896), 1158.
79. Rubinstein, 'Cycling in the 1890s', p. 63.
80. MRC, MSS 328/C/5/HUB: Miss C. Everett-Green, 'Cycling for Ladies', *The Hub* (9 April 1898), 383–4, quote on p. 384.
81. Dora de la Blaquière, 'The Dress for Bicycling', *GOP*, XVII: 823 (5 October 1895), 12–14, quote on p. 12.
82. Mrs Humphrey, 'Women on Wheels', *The Idler* (August 1895), 71–4, quote on p. 71.
83. N.G. Bacon, 'Our Girls A-Wheel', IV, 'Girls' Cycling Clubs and Associations', *GOP*, XVIII: 888 (2 January 1897), 220–1, quote on p. 220.
84. Ross McKibbin, *Classes and Cultures: England, 1918–1951* (Oxford: Oxford University Press, 1998), p. 379.
85. Everett-Green, 'Cycling for Ladies', p. 384.
86. T.P.W., 'The Cycling Epidemic', *The Scottish Review* (29 January 1897), 56–74; Whorton, 'The Hygiene of the Wheel', p. 69.
87. Editorial, 'The Contemporary Review', *Review of Reviews* (May 1898), 480–1, quote on p. 481.

88. Whorton, 'The Hygiene of the Wheel', p. 79.
89. Robert Dickinson, 'Bicycling for Women from the Standpoint of the Gyne-cologist', *American Journal of Obstetrics*, 31 (1895), 24–35. Cited in Whorton, 'The Hygiene of the Wheel', p. 81.
90. 'Abstracts from Foreign Journals', R.L. Dickinson (*American Journal of Obstetrics*, January 1895), *Practitioner*, LIV (March 1895), 280.
91. A.T. Schofield, 'The Cycling Craze', *GOP*, XVII: 834 (21 December 1895), 185–6, quotes on p. 186.
92. Harmond, 'Progress and Flight', p. 244.
93. MRC, MSS.328/C/4/CYT: Robt N. Ingle, 'Cycling for the Young' (Correspondence), *Cyclists' Touring Club Monthly Gazette and Official Record*, XIV (7 July 1895), 200.
94. F.L. Gerald, 'The Bicycle Craze', *Good Health*, VI: 149 (10 August 1895), 577; Mary Lynn Stewart, *For Health and Beauty: Physical Culture for Frenchwomen, 1880–1930s* (Baltimore, MD and London: Johns Hopkins University Press, 2001), p. 84. The link between female cycling and the sewing machine was rebuffed in a feature in *Health* in June 1895, which argued that a much wider range of muscles were employed in cycling and that the association between ill health and the sewing machine was also linked to 'the melancholy conditions amid which it is employed': Dr Just Championniere, 'Woman and the Wheel', *Health* (28 June 1895), 198–9, (5 July 1895), 214–15, on p. 198.
95. Shadwell, 'The Hidden Dangers of Cycling', p. 788.
96. MRC, MSS.328/C/4/CYT: Correspondence 'Cycling for Ladies', Frances Elizabeth Hoggan, *Cyclists' Touring Club Gazette* (December 1887), 454.
97. E.B. Turner, 'A Report on Cycling in Health and Disease', IV 'Cycling for Women', *BMJ*, 1 (6 June 1896), 1399.
98. Ibid.
99. MRC, MSS.328/C/4/CYT: Edward Beardon Turner, 'Cycle Riding and Cycle Racing for Women', *Cyclists' Touring Club Monthly Gazette and Official Record*, XV: 3 (March 1896), 94–6, p. 95.
100. Editorial, 'Bicycling for Women', *Lancet*, 1 (16 May 1896), 1369–70, quote on p. 1369.
101. MRC, CV1041G7: Susan, Countess of Malmesbury, G. Lacy Hillier, and H. Graves, *Cycling* (London: Lawrence and Bullen, 1898), Part III 'Cycling for Women', p. 91.
102. Haydn Brown, 'Cycling for Women: Its Effects on Parturition' (Correspondence), *Lancet*, 2 (24 July 1897), 221–2.
103. Championniere, 'Woman and the Wheel', p. 198.
104. MRC, CV1041G7: Countess of Malmesbury, Lacy Hillier, and Graves, *Cycling*, p. 93.
105. Anon., 'Celebrated Nonsense', *Good Health*, V: 115 (15 December 1894), 161.
106. Section of Domestic Hygiene, 'The Sanitary Aspect of Cycling for Ladies', *BMJ*, 2 (12 September 1896), 681.
107. Turner, 'Cycle Riding and Cycle Racing for Women', p. 95.
108. MRC, MSS 328/C/5/HUB: Anon., 'Why Women should Ride', *Hub* (26 September 1896), 312.

109. MRC, MSS.328/C/4/CYT: Heather Bigg, 'Cycling as an Exercise for Women', *Cyclists' Touring Club Monthly Gazette and Official Record*, XV: 5 (May 1896), 206–8, quote on p. 206.

110. W.H. Fenton, 'A Medical View of Cycling for Ladies', *The Nineteenth Century* (January–June 1896), 796–801, quote on p. 797.

111. Ibid., p. 801.

112. Editor, 'In Praise of Cycling. For Men and Women – Especially Women', *The Review of Reviews* (June 1895), 536–7, quote on p. 536 (his emphasis).

113. Lady Jeune, 'Cycling for Women', *The Badminton Magazine of Sports and Pastimes*, 1: 3 (October 1895), 407–14, quote on p. 413.

114. Anon., 'Cycling for Women: Dr. Sir B.W. Richardson's Views', *Health* (15 November 1895), p. 106; Editorial, 'Cycling for Health and Pleasure', *Chambers's Journal*, XII: 608 (24 August 1895), 529–31, on p. 530.

115. Anon., 'Cycling for Women. Dr. Sir B.W. Richardson's Views'.

116. Fenton, 'A Medical View of Cycling for Ladies', p. 801.

117. Anon., 'British Medical Journal Reports: A Report on Cycling in Health and Disease', p. 1158.

118. Gordon Stables, *The Girl's Own Book of Health and Beauty* (London: Jarrold and Sons, 1891), pp. 89–96. He also published two books devoted to health and cycling: *Rota Vitae: The Cyclist's Guide to Health and Rational Enjoyment* (London: Iliffe & Son, 1886); *Health upon Wheels; or, Cycling a Means of Maintaining the Health* (London: Iliffe & Son, 1887).

119. Stables, *Health upon Wheels*, p. 43.

120. Ibid., p. 45.

121. Richardson, 'On Recreation for Girls', p. 546.

122. N.G. Bacon, 'Our Girls A-Wheel', III 'The Advantages and Pleasures of the Pastime', *GOP*, XVIII: 886 (19 December 1896), 182–3, quote on p. 182.

123. Lawrence Liston, 'Bicycling to Health and Fortune', Part I, 'The Machine', *GOP*, XIX: 930 (23 October 1897), 52–3, quote on p. 52; Part II, 'The Rider', *GOP*, XIX: 939 (25 December 1897), 198–9.

124. Ibid., pp. 52–3.

125. Anon., 'Cycling as a Therapeutic Agent', *BMJ*, 1 (2 February 1889), 252–3. Cited in Bartrip, *Mirror of Medicine*, p. 151.

126. MRC, MSS 328/C/5/HUB: Anon., 'Bicycles on the Brain. People who go Crazy over Cycling', *Hub* (5 February 1898), 33–4, on p. 34; Humphreys, *Personal Hygiene for Girls*, p. 120.

127. T.P.W., 'The Cycling Epidemic', p. 60.

128. Editorial, 'The Contemporary Review', *Review of Reviews* (May 1898), 480–1.

129. De la Blaquière, 'The Dress for Bicycling', pp. 12–13.

130. Editorial, *The Contemporary Review* (May 1898), 481; E.B. Turner, 'Health on the Bicycle', *The Contemporary Review*, (May 1898), 640–48, quote on p. 643.

131. Fenton, 'A Medical View of Cycling for Ladies', pp. 799–800.

132. MRC CV1041G7: Countess of Malmesbury, Lacy Hillier, and Graves, *Cycling*, p. 93.

133. *Forget-Me-Not*, 1: 26 (1892), 14. Cited in Mitchell, *The New Girl*, p. 110.

134. Mitchell, *The New Girl*, p. 110.

135. T.P.W., 'The Cycling Epidemic', p. 68.

136. MRC, MSS 328/C/5/HUB: Anon., 'Why Women should Ride', p. 312.

137. MRC, MSS.328/C/5/BUT: *At the Sign of the Butterfly*, 1897–1898, 'The Editorial Tandem', II: 3 (26 November 1897), 1; Editorial, 'Co-operative Camps for Cyclists: A Suggestion for the Solution of a Social Problem', *The Review of Reviews* (May 1894), 516–21.

138. MRC, MSS.328/C/5/BUT: *At the Sign of the Butterfly*, 1897–1898, 'Rational Dress' [By a Rational Rational-Dresser] (26 November 1897), 14.

139. MRC, MSS.328/C/5/WHE: Anon., 'Outdoor Life for Women', *Cyclists Touring Club. The Wheeler* (16 May 1894), 97.

140. Liston, 'Bicycling to Health and Fortune', Part II, 'The Rider', p. 199.

141. MRC, MSS.328/C/5/BUT: *At the Sign of the Butterfly*, 1897–1898, Edward Bellamy, 'Woman in 2000' (26 November 1897), 5–6.

142. Liston, 'Bicycling to Health and Fortune', Part II, 'The Rider', p. 199.

143. Schofield, 'The Cycling Craze', p. 185.

144. Anson Rabinbach, *The Human Motor: Energy, Fatigue and the Origins of Modernity* (Berkeley, CA: University of California Press, 1992).

145. Fenton, 'A Medical View of Cycling for Ladies', p. 800.

146. Anon., 'Woman on Wheels', *The Speaker* (9 May 1896), 506–7, quote on p. 506.

147. Championniere, 'Woman and the Wheel', p. 214.

148. McCrone, *Sport and the Physical Emancipation of English Women*, pp. 242, 239.

149. MRC, MSS 328/C/5/HUB: Anon., 'A Lady Doctor on Cycling', *Hub* (26 September 1896), 287.

150. Ballin, 'Introduction', pp. 13–14, 15.

151. Bonnell, 'The Power of the Pedal', p. 226.

152. MRC, MSS.328/N28/8/1: Commonplace book of Emily Sophia Coddington, and later cycling diary [c.1850]-1896, entries 7 and 17 February 1895, np. The cycling diary is written in a different hand at the back of the commonplace book.

153. Andrews, 'Making Do'. Cited in Burnett (ed.), *Destiny Obscure*, pp. 125–6.

154. MRC, MSS.328/C/4/CYT: Anon., 'Cycling Dress for Women', *Cyclists' Touring Club Monthly Gazette and Official Record*, XV: 4 (April 1896) 167–8, on p. 167.

155. Miss Lillias Campbell Davidson, 'Cycling', in Whitley, *Every Girl's Book of Sport, Occupation and Pastime*, p. 187.

156. McCrone, *Sport and the Physical Emancipation of English Women*, pp. 180–1.

157. Graves, Hillier, and Countess of Malmesbury, *Cycling*, p. 96; Liston, 'Bicycling to Health and Fortune', Part I, p. 52.

158. MRC, CV 1041 G7: H. Hewitt Griffin, *Cycles and Cycling. With a Chapter for Ladies*, by Miss L.C. Davidson (London: George Bell & Sons, 1890), p. 88. See Jill Matthews, 'They had Such a Lot of Fun: The Women's League of Health and Beauty Between the Wars', *History Workshop*, 30 (1990), 22–54.

159. Bacon, 'Our Girls A-Wheel', III, p. 182. See also Harvey Taylor, *A Claim on the Countryside: A History of the British Outdoor Movement* (Edinburgh: Keele University Press, 1997).

160. Dr W.F. Prather, 'On the Bicycle: The Correct and Healthful Position', *Good Health*, IV: 97 (11 August 1894), 253.

161. A.B. Barnard, *The Girl's Book About Herself* (London: Cassell, 1912), p. 9; Ballin, 'Introduction', p. 8.

162. Marianne Farningham, *Girlhood*, rev. edn (London: James Clarke, 1895), p. 174.
163. Marianne Farningham, *Girlhood* (London: James Clarke, 1869), p. 21; rev. edn (1895), pp. 174–5. See also Mitchell, *The New Girl*, p. 109.
164. Editor of 'Physical Culture', 'The Physical Training of Girls', in Spicer (ed.), *Sports for Girls*, pp. 74–86, quote on p. 76.
165. Lawrence James, *Rise and Fall of the British Empire* (London: Little Brown, 1994), p. 320. Cited in Jane Potter, *Boys in Khaki, Girls in Print: Women's Literary Responses to the Great War 1914–1918* (Oxford: Clarendon, 2005), pp. 36–7.
166. Ballin, 'Introduction', p. 7.
167. Barnard, *The Girl's Book About Herself*, p. 18.
168. Gordon Stables ('Medicus'), 'Health', *GOP*, XXII: 1128 (10 August 1901), 716–17, quote on p. 716.
169. Gordon Stables ('Medicus'), 'Man-Games That Murder Beauty', *GOP*, XXVII: 1376 (12 May 1906), 502–3, on p. 503.
170. Ibid.
171. Mrs Eric Pritchard, 'Physical Culture for Women', *Pall Mall Magazine*, 33 (May 1904), 141–4, quote on p. 141. For this pendulum swing in the US, see Gregory Kent Stanley, 'Redefining Health: The Rise and Fall of the Sportswoman. A Survey of Health and Fitness Advice for Women, 1860–1940', unpublished University of Kentucky PhD thesis, 1991, Chapter 6.

4 Girls, Education and the School as a Site of Health

1. North London Collegiate School Archives (NLCSA): Mrs Hoggan's private notes, RS 1i, 29 September 1882, 5 October 1882, 22 February 1884.
2. Ibid., RS 1i, 8 February 1883, 14 June 1883, 9 May 1885.
3. The first novel to capitalise on the 'Girton Girl' was Annie Edwardes, *A Girton Girl*, serialised in *Temple Bar* in 1881 and published in 1885. See Sally Mitchell, *The New Girl: Girls' Culture in England 1880–1915* (New York: Columbia University Press, 1995), Chapters 3 and 4, for popular fiction based on female college graduates and schoolgirls. This chapter will not examine the health of female university students, though this is touched on briefly in connection with sport in Carol Dyhouse, *No Distinction of Sex? Women in British Universities, 1870–1939* (London: UCL Press, 1995). See, for a discussion of health and body image in the US, Margaret A. Lowe, *Looking Good: College Women and Body Image, 1875–1930* (Baltimore, MD and London: Johns Hopkins University Press, 2003).
4. For sport at girls' schools and colleges, see Paul Atkinson, 'Fitness, Feminism and Schooling', in Sara Delamont and Lorna Duffin (eds), *The Nineteenth-Century Woman: Her Cultural and Physical World* (London: Croom Helm, 1978), pp. 92–133; Kathleen E. McCrone, 'Play Up! Play Up! And Play the Game! Sport at the Late Victorian Girls' Public School', *Journal of British Studies*, 23 (1984), 106–34; idem, *Sport and the Physical Emancipation of English Women 1870–1914* (London: Routledge, 1988); Sheila Fletcher, *Women First: The Female Tradition in English Physical Education 1880–1980* (London: Athlone, 1984); Dyhouse, *No Distinction of Sex?*, pp. 201–6 and

for the US Patricia A. Vertinsky, *The Eternally Wounded Woman: Women, Doctors, and Exercise in the Late Nineteenth Century* (Urbana and Chicago, IL: University of Illinois Press, 1994; first published Manchester: Manchester University Press, 1989); *idem*, 'Body Shapes: The Role of the Medical Establishment in Informing Female Exercise and Physical Education in Nineteenth-century North America', in J.A. Mangan and Roberta J. Parke (eds), *From 'Fair Sex' to Feminism: Sport and the Socialization of Women in the Industrial and Post-Industrial Eras* (London: Frank Cass, 1987), pp. 256–81; Martha H. Verbrugge, *Active Bodies: A History of Women's Physical Education in Twentieth-Century America* (Oxford and New York: Oxford University Press, 2012).

5. Atkinson, 'Fitness, Feminism and Schooling', p. 92.
6. 1887 [C.5158] Elementary Education Acts. Third Report of the Royal Commission Appointed to Inquire into the Working of the Elementary Education Acts, England and Wales, Evidence of Mdme Bergman-Österberg, columns 52, 154–5: http://0-gateway.proquest.com.pugwash.lib.warwick.ac.uk/:hcpp:fulltext:1887-063419:391.
7. Ibid., column, 52, 204–5: http://0-gateway.proquest.com.pugwash.lib.warwick.ac.uk/:hcpp:fulltext:1887-063419:394.
8. Ibid., column 52,163: http://0-gateway.proquest.com.pugwash.lib.warwick.ac.uk/:hcpp:fulltext:1887-063419:392; Mrs Ely-Dallas, Organising Teacher of Physical Exercises under the School Board for London, 'Physical Education in the Girls' and Infants' Departments in the Schools of the London School Board', in *Special Reports on Educational Subjects*, vol. 2 (London: Her Majesty's Stationary Office, 1898), pp. 202, 204–5.
9. Sheila Fletcher, 'Österberg, Martina Sofia Helena Bergman (1849–1915)', *Oxford Dictionary of National Biography* (hereafter *DNB*): [http://0-www.oxforddnb.com.pugwash.lib.warwick.ac.uk/view/article/47656, accessed 27 April 2012], *idem, Women First*, especially pp. 20–41.
10. See, for example, K. Bathhurst, 'The Physique of Girls', *Nineteenth Century and After* (May 1906), 825–33.
11. Carolyn Steedman, 'Bodies, Figures and Physiology: Margaret McMillan and the Late Nineteenth-Century Remaking of Working-Class Childhood', in Roger Cooter (ed.), *In the Name of the Child: Health and Welfare 1880–1940* (London and New York: Routledge, 1992), pp. 19–44, on p. 34; *idem, Childhood, Culture and Class in Britain: Margaret McMillan, 1860–1931* (London: Virago, 1990), Chapter 4.
12. Peter C. McIntosh, *Physical Education in England since 1800*, revised edn (London: G. Bell & Sons, 1968), p. 158. For the School Medical Service, see Bernard Harris, *The Health of the Schoolchild: A History of the School Medical Service in England and Wales* (Buckingham and Philadelphia, PA: Open University Press, 1995); Harry Hendrick, 'Child Labour, Medical Capital, and the School Medical Service, c.1890–1918', in Cooter (ed.), *In the Name of the Child*, pp. 45–71, and for physical education and ideas on citizenship in Britain, John Welshman, 'Child Health, National Fitness, and Physical Education in Britain, 1900–1940', in Marijke Gijswijt-Hofstra and Hilary Marland (eds), *Cultures of Child Health in Britain and the Netherlands in the Twentieth Century* (Amsterdam and New York: Rodopi, 2003), pp. 61–84.

13. Medical Officer of Health Reports, London, 1906, Appendix 2 (Report of Medical Officer (Education)), 1908, Appendix 5. Cited in Anna Davin, *Growing Up Poor: Home, School and Street in London 1870–1914* (London: Rivers Oram Press, 1996), pp. 77–8.

14. Miss Chreiman, *Physical Culture of Women: Lecture, by Request of the Council of Parkes' Museum, March 15th, 1888* (London: Sampson Low, Marston, Searle, & Rivington, 1888), pp. 8–9; see also *idem, Physical Education of Girls. A Lecture Delivered in the Lecture Room of the Exhibition, July 25th, 1884*, Published for the Executive Council of the International Health Exhibition and the Council of the Society of Arts (London: William Clowes and Sons, 1884).

15. Chreiman, *Physical Culture of Women*, pp. 34–5.

16. Ibid., p. 6.

17. Chreiman, *Physical Education of Girls.*

18. See M.A. Elston, 'Hoggan [*née* Morgan], Frances Elizabeth (1843–1927)', *DNB*: [http://0-www.oxforddnb.com.pugwash.lib.warwick.ac.uk/view/article/46422, accessed 7 February 2012].

19. NLCSA: History of the School 1875–1894 B1, Pamphlet 'Physical Education' advertising Miss Chreiman's classes with testimonials (n.d., c.1883).

20. Anon., 'Physical Education of Women', *Health* (4 January 1884), 217.

21. Frances Elizabeth Hoggan, *On the Physical Education of Girls, Read at the Annual Meeting of the Fröbel Society*, on December 9, 1879 (London: W. Swan Sonnenschein & Allen, 1880), pp. 10–11.

22. Ibid., p. 10.

23. Hoggan took up the poetic form in the popular periodical *Health*:
    'Rounded shoulders, slouching gait, and also haply crooked spines,
    By gymnastic exercises shall grow straight as mountain pines;
    Let the girls then learn athletics who in town are apt to droop,
    Careful drill will make them upright and eradicate the stoop.'
    [Frances Hoggan], 'Girl Gymnasts', *Health* (21 March 1884), 392.

24. Chreiman, *Physical Education of Girls*, p. 18.

25. Chreiman, *Physical Culture of Women*, p. 31.

26. A. Alexander, *Healthful Exercises for Girls* (London: George Philip & Son, 1886), p. iv.

27. Ibid., pp. 22, 64, 60.

28. Clement Dukes, *Health at School Considered in its Mental, Moral, and Physical Aspects* (London, Paris, New York & Melbourne: Cassell, 1887), pp. 311–16.

29. Clement Dukes, 'Health at School', in Malcolm Morris (ed.), *The Book of Health* (London, Paris and New York: Cassell, 1883), pp. 677–725, quotes on p. 725.

30. Jane Frances Dove, 'Cultivation of the Body', in Dorothea Beale, Lucy H.M. Soulsby and Jane Frances Dove, *Work and Play in Girls' Schools by Three Head Mistresses* (London, New York and Bombay: Longmans, Green, and Co., 1898), pp. 396–423, quote on p. 410.

31. NLCSA: History of the School 1875–1894 (1), B1 Appointment of Doctor.

32. NLCSA: Gymnastic Medical Notes, 1882–6, Miss Hoggan's Notes 1882–5. Cited in McCrone, *Sport and the Physical Emancipation of English Women*, p. 64; Josephine Kamm, *How Different from Us – Miss Buss and Miss Beale* (London: Bodley Head, 1958), p. 224.

33. NLCSA: Mrs Hoggan's private notes, RS 1i, 14 June, 5 October 1883.
34. Ibid., 5 October 1882.
35. NLCSA: Gymnasium Medical Notes, RS 1 (ii), 29 September 1882, 5 June 1884.
36. NLCSA: 'In Memoriam', *North London Collegiate School for Girls, Our Magazine*, 1882–88: April 1882, 133.
37. Henry Maudsley, 'Sex in Mind and in Education', *Fortnightly Review*, new series, 15 (April 1874), 466–83, quotes on pp. 467, 477. See also Joan N. Burnstyn, 'Education and Sex: The Medical Case against Higher Education for Women in England, 1870–1900', *Proceedings of the American Philosophical Society*, 117 (1973), 79–89; Anne Digby, 'Women's Biological Straitjacket', in Susan Mendus and Jane Rendall (eds), *Sexuality and Subordination: Interdisciplinary Studies of Gender in the Nineteenth Century* (London and New York: Routledge, 1989), pp. 192–200, especially pp. 208–14.
38. Maudsley, 'Sex in Mind and in Education', p. 466.
39. See Jane Martin, *Women and the Politics of Schooling in Victorian and Edwardian England* (London and New York: Leicester University Press, 1999), Chapter 4 for women's participation in educational politics.
40. Elizabeth Garrett Anderson, 'Sex in Mind and Education: A Reply', *Fortnightly Review*, new series, 15 (May 1874), 582–94, quotes on pp. 585, 586.
41. Ibid.; *idem*, 'Examinations for Women', *The Times* (12 November 1870), 10 and Jo Manton, *Elizabeth Garrett Anderson* (London: Methuen, 1965).
42. Joyce Senders Pederson, 'The Reform of Women's Secondary and Higher Education: Institutional Change and Social Values in Mid and Late Victorian England', *History of Education Quarterly*, 19 (Spring 1979), 61–91, on p. 73; June Purvis, *A History of Women's Education in England* (Milton Keynes and Philadelphia, PA: Open University Press, 1991), p. 76. For girls' schooling, see also Felicity Hunt (ed.), *The Schooling of Girls and Women, 1850–1950* (Oxford: Basil Blackwell, 1987); Joan Burstyn, *Victorian Education and the Ideal of Womanhood* (London: Croom Helm, 1980); Carol Dyhouse, *Girls Growing up in Late Victorian and Edwardian England* (London: Routledge & Kegan Paul, 1981).
43. Sir Alexander Grant, *Happiness & Utility as Promoted by the Higher Education of Women* (Edinburgh: Edmonston & Douglas, 1872), p. 18. Cited in Burstyn, 'Education and Sex', p. 82.
44. 'Reports issued by the Schools' Inquiry Commission, on the Education of Girls. With Extracts from the Evidence and a Preface (Taunton Report)', by D. Beale, Principle of the Ladies' College Cheltenham (London: David Nutt, 1869), Report of Mr Fitch, Mr Bryce's Report, pp. 41–2, 65–6.
45. Ibid., Mr Fearon's Report, p. 85.
46. Ibid., Miss Wolstenholme's Evidence, p. 232.
47. Ibid., Dorothea Beale's Evidence, pp. 206–7.
48. McCrone, 'Play Up! Play Up!', pp. 112–13.
49. Barbara Stephen, *Emily Davies and Girton College* (London: Constable, 1927), p. 95.
50. Elizabeth Blackwell, *Lectures on the Laws of Life With Special Reference to the Physical Education of Girls* (London: Sampson, Low, Son, & Marsten, 1871), p. 129.
51. Ibid., p. 172.

52. Elizabeth Coutts, 'Buss, Frances Mary (1827–1894)', *DNB* [http://0-www. oxforddnb.com.pugwash.lib.warwick.ac.uk/view/article/37249, accessed 25 April 2012].

53. T.S. Clouston, 'Puberty and Adolescence Medico-Psychologically Considered', *Edinburgh Medical Journal*, 26: 1 (July 1880), 5–17, quotes on pp. 8, 10.

54. T.S. Clouston, 'Female Education from a Medical Point of View', *Popular Science Monthly*, 24 (December 1883, January 1884), 214–28, 319–34, quote on p. 325. See also the extracts in Pat Jalland and John Hooper (eds), *Women from Birth to Death: The Female Life Cycle in Britain 1830–1914* (Brighton: Harvester, 1986), pp. 77–86.

55. John Thorburn, *Female Education in its Physiological Aspect* (Manchester: J.E. Cornish, 1884).

56. Idem., *A Practical Treatise on the Diseases of Women* (London: Charles Griffin, 1885), p. 99.

57. Ibid., pp. 99–100. Few British medical writers considered themselves conservative in their attitudes to women. Robert Lawson Tait declared himself in favour of women's rights and while serving at the Birmingham Women's Hospital promoted medical women's interests by appointing them to hospital posts. At the same time, he believed that women should not avail themselves too fully of those rights. See Judith Lockhart, 'Women, Health and Hospitals in Birmingham: The Birmingham and Midland Hospital for Women, 1871–1948', unpublished University of Warwick PhD thesis 2008, especially Chapters 2 and 3, for Tait's support of women doctors in Birmingham.

58. M. Tylecote, *The Education of Women at Manchester University, 1883–1933* (Publications of the University of Manchester, no. 277, 1941). Cited in Dyhouse, *Girls Growing Up*, p. 155.

59. E.J. Tilt, *Elements of Health, and Principles of Female Hygiene* (London: Henry G. Bohn, 1852), pp. 150–6.

60. Edward John Tilt, 'Education of Girls', *British Medical Journal* (*BMJ*), 1 (2 April 1887), 751; Thorburn, *Female Education*, p. 8.

61. Thorburn, *Female Education*, pp. 8–9.

62. Ibid., p. 15.

63. Howard A. Kelly, *Medical Gynecology* (London: Appleton, 1908), p. 66. This was also affirmed by the Medical Women's Federation from the 1920s onwards: Julie-Marie Strange has explored menstrual hygiene, including approaches in girls' schools, in 'The Assault on Ignorance: Teaching Menstrual Etiquette in England, c.1920s to 1960s', *Social History of Medicine*, 14 (2001), 247–65.

64. Emily Pfeiffer, *Women and Work: An Essay Treating on the Relation to Health and Physical Development, of the Higher Education of Girls, and the Intellectual or More Systematised Effort of Women* (London: Trüber & Co., 1888).

65. Response of William N. Gull, in Pfeiffer, *Women and Work*, p. 92.

66. Letter from Hermann Weber, dated December 1886, in Pfeiffer, *Women and Work*, p. 93.

67. W.S. Playfair, 'Remarks on the Education and Training of Girls of the Easy Classes at and about the Period of Puberty', *BMJ*, 2 (7 December 1895), 1408–10, p. 1408.

68. Ibid., p. 1409.

69. Ibid.
70. Ibid., pp. 1409–10.
71. Kate Flint, *The Woman Reader 1837–1914* (Oxford: Clarendon, 1993), p. 57.
72. Anon., 'Education of Girls in America', *Good Health*, IV: 84 (12 May 1894), 87.
73. Jennie Chandler, 'Hygiene for Women. A New Education for Women', *Good Health*, IV: 91 (30 June 1894), 187.
74. Anon., 'Woman's Beauty', *Good Health*, VIII: 188 (9 May 1896), 170–1.
75. Ibid., p. 171.
76. Ibid.
77. J. Crichton-Browne, 'Education and the Nervous System', in Morris, *The Book of Health*, pp. 269–380, on p. 312.
78. Sir James Crichton-Browne, 'An Oration on Sex in Education. Delivered at the Medical Society of London on May 2nd, 1892', *Lancet*, 1 (7 May 1892), 1011–18, quotes on pp. 1014, 1013.
79. Clement Dukes, 'The Hygiene of Youth', in T.C. Allbutt and H.D. Rolleston (eds), *A System of Medicine by Many Writers*, vol. 1 (London and New York: Macmillan, 1905), pp. 160–81, on p. 179.
80. Dukes, 'Health at School', p. 724.
81. For example, Dukes, *Health at School*; idem, *Work and Overwork in Relation to Health in Schools*, An Address Delivered before the Teachers' Guild of Great Britain and Ireland at its Fifth General Conference Held in Oxford, 17–20 April 1893 (London: Percival, 1893), p. 48.
82. Dukes, *Health at School*, pp. 307–8.
83. Clement Dukes, *The Essentials of School Diet or the Diet Suitable for the Growth and Development of Youth*, 2nd edn (London: Rivingtons, 1899).
84. D. Beale, *On the Education of Girls: A Paper Read at the Social Science Congress, October, 1865, and reprinted from the Transactions* (London: Bell & Daldy, 1866).
85. Ibid., p. 14.
86. Mrs S. Bryant, *Over-work: From the Teacher's Point of View with Special Reference to the Work in Schools for Girls*, A Lecture Delivered at the College of Preceptors, November 19th, 1884 (London: Francis Hodgson, 1885), p. 3.
87. Ibid., pp. 5–9.
88. Ibid. p. 16.
89. Atkinson, 'Fitness, Feminism and Schooling'.
90. That the impact of schooling on girls' health was actively engaged with by Frances Buss is evidenced by the small collection of her books and pamphlets held in the Library of the NLCS. This includes two pamphlets on the health statistics of female students at Oxford and Cambridge, compiled by Mrs Henry Sidgwick and published in 1890, and on the health of schools in America, produced in 1876 by the American Social Science Association, which Buss had carefully annotated; both texts pointed to the positive impact of schools on the health of young women: Mrs Henry Sidgwick, *Health Statistics of Women Students of Cambridge and Oxford and of their Sisters* (Cambridge: Cambridge University Press, 1890); *The Health of Schools: Papers Read before the American Social Science Association, at Detroit, May 1875* (1876), NLCSA: RS7iV Personal Copy FMB, with signature.
91. Fletcher, *Women First*, p. 35.

92. NLCSA: RS 1(i), Minutes 1875–1889, 7 July and 1 December 1879, pp. 190, 199.
93. Anne E. Ridley, *Frances Mary Buss and Her Work for Education* (London: Longmans, Green and Co, 1895), p. 222.
94. Dyhouse, *Girls Growing Up*, p. 154; McCrone, *Sport and the Physical Emancipation of English Women*, p. 67.
95. R.M. Scrimgeour (ed.), *The North London Collegiate School 1850–1950: A Hundred Years of Girls' Education* (London, New York and Sydney: Oxford University Press, 1950), p. 46.
96. Dorothea Beale, *Address to Parents* (London: George Bell & Sons, 1888), pp. 7–8.
97. Harris, *The Health of the Schoolchild*, p. 207.
98. Kelly, *Medical Gynecology*, pp. 52, 69. Kelly also cited figures from New York schools where 63 per cent of children entering school needed medical treatment, 'a tremendous indictment against the efficiency of the home' (p. 55).
99. Ibid., p. 62.
100. Playfair, 'Remarks on the Education and Training of Girls', p. 1409.
101. Edred M. Corner, 'Physical Exercises' in Allbutt and Rolleston (eds), *A System of Medicine by Many Writers*, vol. 1, pp. 382–421, quote on p. 395.
102. McCrone, 'Play Up! Play Up!', p. 118.
103. *Cheltenham Ladies College Magazine* (Spring 1893), p. 123 (Spring 1895), p. 182. Cited Ibid., p. 119; F. Cecily Steadman, *In the Days of Miss Beale: A Study of Her Work and Influence* (London: E. J. Burrow, 1936), p. 11.
104. Modern Records Centre (MRC), MSS.188/3/6/1 Papers Relating to the History of the Association of Head Mistresses 1874, 1927–1962, Transcript 'Woman's Hour' 'Dorothea Beale', 8 November 1956.
105. Report of the Lady Principle for 1893–94, *Cheltenham Ladies' College Magazine* (Spring 1895), p. 182. Cited in McCrone, *Sport and the Physical Emancipation of English Women*, p. 83.
106. Jacqueline Beaumont, 'Beale, Dorothea (1831–1906)', *DNB*: [http://www.oxforddnb.com.pugwash.lib.warwick.ac.uk/view/article/30655, accessed 25 April 2012].
107. MRC, MSS.59/4/1/1 Annual Reports: Association of Assistant Mistresses. Report of the Fourth Year's Work of the Association, and of the Fourth Annual Meeting, January 1888, p. 16.
108. MRC, MSS. 188/4/1/1 Annual Reports: Association of Headmistresses 1895–1900, Report 1895, p. 7.
109. Dove, 'Cultivation of the Body', p. 397.
110. Ibid., pp. 398–9.
111. Ibid., pp. 399–400.
112. Mary C. Malim and Henrietta C. Escreet (eds), *The Book of Blackheath High School* (London: Blackheath Press, 1927), pp. 109–11, 116. Cited in McCrone, *Sport and the Physical Emancipation of English Women*, p. 69. See also *idem.*, 'Play Up!, Play Up'.
113. Dove, 'Cultivation of the Body', p. 423.
114. Ibid., pp. 412, 414.
115. Steadman, *In the Days of Miss Beale*, p. 70.

116. Sara Burstall, *English High Schools for Girls: Their Aims, Organisation, and Management* (London, New York, Bombay, and Calcutta: Longmans, Green, and Co., 1907), p. 96. For sporting attire at school, see McCrone, *Sport and the Physical Emancipation of English Women*, Chapter 8, especially pp. 224–8.

117. Burstall, *English High Schools for Girls*, p. 97.

118. Ibid., Chapter XIII.

119. See Davin, *Growing Up Poor*, especially pp. 142–9; Carol Dyhouse, 'Good Wives and Little Mothers: Social Anxieties and the Schoolgirl's Curriculum, 1890–1920', *Oxford Review of Education*, 3 (1977), 21–35; Ellen Jordan, ' "Making Good Wives and Mothers"? The Transformation of the Middle-Class Girls' Education in Nineteenth-Century Britain', *History of Education Quarterly*, 31 (1991), 439–62.

120. Janet Campbell, 'The Effect of Adolescence on the Brain of the Girl', paper presented to the Association of University Women Teachers meeting, London, 23 May 1908, pp. 5–6. Cited in Dyhouse, *Girls Growing Up*, p. 134.

121. Burstall, *English High Schools for Girls*, p. 194. See Michelle J. Smith, *Empire in British Girls' Literature and Culture: Imperial Girls, 1880–1915* (Houndmills: Palgrave Macmillan, 2011), pp. 64–7 for links between domestic science and Empire.

122. MRC, Archive of the Association of Teachers of Domestic Science, MSS. 177/1/1/2ii, Association of Teachers of Domestic Subjects, May 1911 'Suggestions on the "Teaching of Housecraft" in Secondary Schools offered by the Special Committee Appointed to Consider the Subject' (pamphlet), p. 1.

123. E.S. Chesser, *Physiology and Hygiene for Girls' Schools and Colleges* (London: G. Bell and Sons, 1914).

124. Ibid., preface.

125. Smith, *Empire in British Girls' Literature and Culture*, p. 65.

126. McCrone, 'Play Up!, Play Up', p. 120.

127. Anon., 'Are Athletics Over-done in Girls' Colleges and Schools? A Symposium', *Woman at Home*, 6 (April 1912), 247–52, on pp. 248–9.

128. Ibid.

129. Ibid., p. 251.

130. Catherine Chisholm, *The Medical Inspection of Girls in Secondary Schools* (London, New York, Bombay and Calcutta: Longmans, Green, and Co., 1914), preface, quote on p. viii. Dr Catherine Chisholm was school medical officer at the Manchester Girls High School between 1908 and 1945. See Peter Mohr, 'Women-run Hospitals in Britain: A Historical Survey focusing on Dr Catherine Chisholm (1878–1952) and The Manchester Babies' Hospital (Duchess of York Hospital)', unpublished University of Manchester PhD thesis, 1995 and *idem*, 'Chisholm, Catherine (1878–1952)', *DNB*: [http://www.oxforddnb.com.pugwash.lib.warwick.ac.uk/view/article/46/46395, accessed 7 February 2012].

131. Anon., 'Are Athletics Over-done?', pp. 250–1; Mrs M. Scharlieb, 'Recreational Activities of Girls during Adolescence', *Child-Study: The Journal of the Child-Study Society*, 4 (1911), 1–14.

132. Scharlieb, 'Recreational Activities of Girls during Adolescence', pp. 8, 9.

133. Elizabeth Sloan Chesser, *From Girlhood to Womanhood* (London, New York, Toronto and Melbourne: Cassell, 1913), pp. 45, 46.

134. C. Cowdroy, 'The Danger of Athletics for Girls and Women', *Lancet*, 1 (14 May 1921), 1050. See also [By our Medical Correspondent], 'Boys Games bad for Girls: Ill-Health and Ill-Temper', *Times* (12 April 1921), 7.
135. Margaret G. Thackrah, 'The Danger of Athletics for Girls', *Lancet*, 1 (18 June 1921), 1328. See Fletcher, *Women First*, pp. 74–7 for more details on this debate.
136. Editorial, 'The Physical Education of Girls', *BMJ*, 2 (19 August 1922), 321.
137. Ibid.
138. Chisholm, *The Medical Inspection of Girls in Secondary Schools*, pp. 103, 104, 105.
139. Chesser, *Physiology and Hygiene for Girls' Schools and Colleges*, p. 86.
140. Wellcome Library: SA/MWF/B.2/1/3: Box 12, *The Medical Women's Federation Newsletter* July 1925 – Mar 1927, pp. 53–4: *Advice Regarding Menstruation to Parents, Schoolmistresses and Others in Charge of Girls*. See also Strange, 'Teaching Menstrual Etiquette'.
141. Christine M. Murrell, *Womanhood and Health* (London: Mills & Boon, 1923), p. 96.
142. Alice E. Sanderson Clow, 'Menstruation during School Life', *BMJ*, 2 (2 October 1920), 511–13, quote on p. 513. See also R.W. Johnstone, J.H.P. Paton and Alice E. Sanderson Clow, 'Discussion on the Hygiene of Menstruation in Adolescents', *BMJ*, 2 (10 September 1927), 442–8.
143. M.E. Findlay, 'The Education of Girls', *The Paidologist*, VII (1905), 83–93, quote on p. 87.
144. Scharlieb, 'Recreational Activities of Girls during Adolescence', pp. 12–13.
145. Elizabeth Sloan Chesser, *Perfect Health for Women and Children* (London: Methuen, 1912), pp. 42–3.
146. Chesser, *From Girlhood to Womanhood*, pp. 77, 80.
147. Robert Jones, 'Girls' Schools, Games, and Neurasthenia', *Lancet*, 1 (4 February 1911), 329.
148. Ibid.
149. V. Sturge, 'The Physical Education of Women' in Dale Spender (ed.), *The Education Papers: Women's Quest for Equality in Britain, 1850–1912* (New York and London: Routledge & Kegan Paul, 1987), pp. 284–94, quote on p. 290
150. Fletcher, *Women First*, p. 31.
151. Christina Gowans Whyte, 'Famous Girls' Schools. IV. Roedean', *The Girl's Realm Annual* (November 1899–October 1990) (London: S.H. Bousfield, 1900), pp. 1060–6, on p. 1060.
152. Sturge, 'The Physical Education of Women', p. 286.
153. NLCSA: 'Our Gymnasium', *North London Collegiate School for Girls, Our Magazine*, 1882–88: March 1885, 26–7.
154. By a Girl Graduate, 'A Sketch of Life at our Great Girl Public Schools', *Health* (11 October 1895), 18.
155. Ibid.

5 The Health of the Factory Girl

1. Modern Records Centre, University of Warwick (MRC): Young Women's Christian Association (YWCA) Reports MSS.243/12/2 1862–68. Miss Paton,

'Factory and Laundry Girls', Report of All-Day Convention, at Exeter Hall, 17 April 1885, p. 39.

2. For example, Meg Gomersall, *Working-Class Girls in Nineteenth-Century England: Life, Work and Schooling* (Houndmills: Macmillan, 1997); Carol Dyhouse, *Girls Growing up in Late Victorian and Edwardian England* (London: Routledge & Kegan Paul, 1981); Jane Martin, *Women and the Politics of Schooling in Victorian and Edwardian England* (London: Leicester University Press, 1999); Selina Todd, *Young Women, Work and Family 1918–50* (Oxford: Oxford University Press, 2005).

3. June Purvis, 'The Double Burden of Class and Gender in the Schooling of Working-Class Girls in Nineteenth-Century England, 1800–1870', in Len Barton and Stephen Walker (eds), *Schools, Teachers and Teaching* (Lewes: Falmer, 1981), pp. 97–116.

4. I.S.L. Loudon, 'Chlorosis, Anaemia, and Anorexia Nervosa', *British Medical Journal*, 281 (20–27 December 1980), 1669–75, on p. 1673; Anna Davin, *Growing Up Poor: Home, School and Street in London 1870–1914* (London: Rivers Oram Press, 1996), pp. 137–40.

5. Barbara Harrison, *Not Only the 'Dangerous Trades': Women's Work and Health in Britain, 1880–1914* (London: Taylor & Francis, 1996); Carolyn Malone, *Women's Bodies and Dangerous Trades in England, 1880–1914* (Woodbridge: Boydell, 2003); Arthur J. McIvor, *A History of Work in Britain, 1880–1950* (Houndmills and New York: Palgrave, 2001), Chapter 7; Gail Braybon, *Women Workers in the First World War* (London: Croom Helm, 1981); Gail Braybon and Penny Summerfield, *Out of the Cage: Women's Experiences in Two World Wars* (London and New York: Pandora, 1987); Deborah Thom, *Nice Girls and Rude Girls: Women Workers in World War I* (London: I.B. Taurus, 1997); Antonia Ineson and Deborah Thom, 'T.N.T. Poisoning and the Employment of Women Workers in the First World War', in Paul Weindling (ed.), *The Social History of Occupational Health* (London: Croom Helm, 1985), pp. 89–107.

6. Barbara Harrison, 'Women and Health', in June Purvis (ed.), *Women's History: Britain, 1850–1945* (London: UCL Press, 1995), pp. 157–92.

7. Malone, *Women's Bodies and Dangerous Trades*.

8. See, for example, Jane Lewis, *The Politics of Motherhood: Child and Maternal Welfare in England 1900–1939* (London: Croom Helm, 1980); Deborah Dwork, *War is Good for Babies and Other Young Children: A History of the Infant and Child Welfare Movement in England 1898–1918* (London and New York: Tavistock, 1989); Anna Davin, 'Imperialism and Motherhood', *History Workshop Journal*, 5 (1978), 9–66; Valerie Fildes, Lara Marks and Hilary Marland (eds), *Women and Children First: International Maternal and Infant Welfare 1870–1945* (London and New York: Routledge, 1992).

9. See Vicky Long and Hilary Marland, 'From Danger and Motherhood to Health and Beauty: Health Advice for the Factory Girl in Early Twentieth-Century Britain', *Twentieth Century British History*, 20 (2009), 454–81, which traces girls' health in industry in the first half of the twentieth century, incorporating analysis of government and factory inspector reports on the health of women and girls in industry.

10. Ellen Ross, *Love and Toil: Motherhood in Outcast London, 1870–1918* (New York and Oxford: Oxford University Press, 1993), Chapter 2. This

was not always the case, however, and some girls demanded a better diet, including one of Elizabeth Roberts' Lancashire respondents: 'I worked at the mill half-time, ... and I come home and m'mother had a herring between us. I said, "Mother, I'm working, I should have a whole one"', and she was duly given a whole herring. Mrs Heron (H2L, b.1889) (Lancaster). Elizabeth Roberts, *A Woman's Place: An Oral History of Working-Class Women 1890–1940* (Oxford: Basil Blackwell, 1984), p. 40.

11. Robert Roberts, *The Classic Slum: Salford Life in the First Quarter of the Century* (Penguin edn, 1973; first published Manchester: University of Manchester Press, 1971), p. 109.
12. Ross, *Love and Toil*, p. 154.
13. 1906 [Cd. 3036] Factories and Workshops. Annual Report of the Chief Inspector of Factories and Workshops for the Year 1905. Reports and Statistics, p. 270: http://0-gateway.proquest.com.pugwash.lib.warwick.ac.uk/o. See also Helen Jones, 'Women Health Workers: The Case of the First Women Factory Inspectors in Britain', *Social History of Medicine*, 1 (1988), 165–81.
14. Vicky Long, *The Rise and Fall of the Healthy Factory: The Politics of Industrial Health in Britain, 1914–60* (Houndmills: Palgrave Macmillan, 2011), p. 58.
15. See Bernard Harris, *The Health of the Schoolchild: A History of the School Medical Service in England and Wales* (Buckingham and Philadelphia, PA: Open University Press, 1995).
16. Sidney Webb and Beatrice Webb, *English Poor Law History. Part II: The Last Hundred Years* (London: Longmans, Green, 1929), pp. 602–3. Cited ibid., p. 64.
17. Roberts, *A Woman's Place*, pp. 30–3; Margery Spring Rice, *Working-Class Wives*, 2nd edn (London: Virago, 1981; first published Penguin 1939), p. 46.
18. Davin, *Growing Up Poor*, pp. 147–8.
19. Lucinda McCray Beier, *For Their Own Good: The Transformation of English Working-Class Health Culture, 1880–1970* (Columbus OH: Ohio State University Press, 2008), pp. 316–31.
20. Penny Tinkler, *Constructing Girlhood. Popular Magazines for Girls Growing Up in England 1920–1950* (London: Taylor & Francis, 1995), p. 45.
21. Kirsten Drotner, *English Children and their Magazines, 1751–1945* (New Haven, CT and London: Yale University Press, 1988), pp. 121–2.
22. Gordon Stables, *The Girl's Own Book of Health and Beauty* (London: Jarrold and Sons, 1891), p. 197.
23. 'Answers to Correspondents', *Girl's Own Paper (GOP)*, II: 40 (20 October 1880), 15. Cited in Drotner, *English Children and their Magazines*, pp. 115–16.
24. 'Medicus', 'Common-Sense Advice for Working Girls', *GOP*, VI: 267 (7 February 1885), 295–6, quote on p. 295.
25. Anon., 'Healthier Lives for Working Girls', *GOP*, VIII: 357 (30 October 1886), 76–9, quote on p. 77.
26. Maud Curwen and Ethel Herbert, *Simple Health Rules and Health Exercises for Busy Women and Girls* (London: Simpkin, Marshall, Hamilton, Kent & Co., 1912).
27. James Cantlie, 'The Influence of Exercise on Health', in Malcolm Morris (ed.), *The Book of Health* (London, Paris and New York: Cassell, 1883), pp. 381–461, quote on p. 399.
28. Ibid., p. 400.

29. Long, *The Rise and Fall of the Healthy Factory*, p. 57.
30. Mary E. Talbot, 'The Hooligans of the Female Sex', Letter to Editor, *Times* (25 October 1901), 6; Miss Mededith Brown, 'The Hooligans of the Female Sex', Letter to Editor, *Times* (29 October 1901), 13.
31. Harrison, *Not Only The 'Dangerous Trades'*, pp. 5–6; Todd, *Young Women, Work and Family 1918–50*, pp. 19–20. Harrison and Todd draw upon the Census Returns for England and Wales. See also Jane Lewis, *Women in England 1870–1950: Sexual Divisions and Social Change* (Brighton: Wheatsheaf, 1984), pp. 148, 152.
32. MRC: YWCA, MSS 243/2/1/7: *A Review YWCA 1922* 'The Girl of To-Day', p. 2; Census of England and Wales (1921), Occupational Tables, p. 4. Cited in Tinkler, *Constructing Girlhood*, p. 27.
33. Roberts, *A Woman's Place*, p. 39.
34. Elizabeth Sloan Chesser, 'Women and Girls in the Factory', *Westminster Review*, 173: 5 (May 1910), 516–19, quote on p. 516.
35. Jan Rutter, 'The Young Women's Christian Association of Great Britain, 1900–1925: An Organisation of Change', unpublished University of Warwick MA Dissertation', 1986, pp. 42–4.
36. Ibid., pp. 47–50.
37. MRC: YWCA MSS.243/2/1/9, *The YWCA. The Movement in 1924. A Review*, p. 2.
38. MRC: YWCA, MSS.243/50/1/1 Memo, 26 August 1925. The Need for the Consideration and Re-Statement of the Functions of the Industrial Law Bureau and its Relation to other Y.W.C.A. Committees.
39. MRC: YWCA, MSS 243/12/4, *Young Women's Christian Association and Institute Union. Report for February 1880* (April 1880), 'Leamington YWCA', p. 84.
40. Ibid., (February 1880), 'Dunstable', p. 39.
41. MRC: YWCA Reports MSS. 243/12/2 1862–86. Miss L. Trotter, 'A Plea for Restaurants', Report of an All-Day Convention, at Exeter Hall, 17 April 1885, p. 15.
42. Ibid., Hon. Emily Kinnaird, 'Social Improvement and Home Comfort', p. 33.
43. Beier, *For Their Own Good*, p. 313.
44. MRC: YWCA, MSS 243/5/11, *Our Own Gazette*, 37 (November 1900), Anon., 'Food Under Difficulties: Hints to Girls who Work', 52–3, quote on p. 53.
45. Ibid., 37 (1900, no issue no.), Kate L.B. Moorhead, 'The Evolution of a Dinner. Being the Story of a Stale Loaf', 77.
46. Ibid., 37 (1900, no issue no.), Anon., 'The Manchester Girls' Institute: A Wonderful Record', 4–6; 'N.O.C.G. Notes', *Girls' Club News*, No. 11 (January 1913), 8.
47. MRC: YWCA, MSS 243/2/1/21, YWCA *'From Strength to Strength'*, Review *1938–39*, p. 2.
48. MRC: YWCA, MSS 243/5/1, *Our Own Gazette*, 1 (September 1884), Anon., 'Tight Lacing: An Important Subject', 105.
49. MRC: YWCA, MSS 243/4/2/1, *Our Own Gazette*, 40 (March 1922), Anon., 'The Care of the Teeth', 14.
50. Girlguiding UK Archives, *Girl Guides' Gazette*, VII: 93 (January 1920), Anon., 'Care of the Teeth', 8.

51. MRC: YWCA, MSS.243/5/1, *Our Own Gazette*, 1 (October 1884), Mr A. Alexander, 'Ladies Gymnasium', 112.
52. MRC: YWCA, MSS.243/5/1, *Our Own Gazette*, 1 (July 1884), Miss Kate Alden, 'Healthful Exercises for Girls', 80.
53. MRC: YWCA, MSS.243/4/3/2, *Our Own Gazette*, 40 (February 1923), Anon., 'To Improve the Figure', 17.
54. MRC: YWCA, MSS.243/4/1/5, *Our Own Gazette*, 38 (May 1920), Mary Lermitte, 'Health and Hygiene', 22.
55. Ibid., (April 1920), Lermitte, 'Health and Hygiene', 22.
56. MRC: YWCA, MSS.243/5/4, *Our Own Gazette*, 10 (September 1893), Ellen La Garde, 'Outdoor Sports for Girls', 106–7.
57. S.F.A. Caulfeild, *A Directory of Girls' Societies, Clubs, and Unions, Conducted on Unprofessional Principles* (London: Griffith, Farran, Okeden & Welsh, 1886), p. 5. Caulfeild also contributed regularly to the *GOP*.
58. London Metropolitan University, The Women's Library (WL), The Girls' Friendly Society (GFS), 1875–5, 5GFS/5/15 'How to Gain the GFS Leaders' Certificate in Physical Activities', p. 5.
59. Captain Baker, 'Drill Brigade for Girls', *Girl's Realm* (September 1908). Cited in Jane Potter, *Boys in Khaki, Girls in Print: Women's Literary Responses to the Great War 1914–1918* (Oxford: Clarendon, 2005), p. 38.
60. Barclay Baron, *The Growing Generation: A Study of Working Boys and Girls in our Cities* (London: Student Christian Movement, 1911), preface, p. vii.
61. Ibid., p. 3.
62. Ibid., pp. 8–9.
63. Agnes Smyth Baden-Powell and Sir Robert Baden-Powell, *The Handbook for Girl Guides; or How Girls Can Help Build the Empire* (London: Thomas Nelson and Sons, 1912), pp. 319–20. See, for an overview of the history of the Girl Guides, Tammy M. Proctor, *Scouting for Girls: A Century of Girl Guides and Girl Scouts* (Santa Barbara, CA: Praeger, 2009), for guiding in the US, Susan A. Miller, *Growing Girls: The Natural Origins of Girls' Organizations in America* (New Brunswick, NJ and London: Rutgers University Press, 2007), and for the link between girls' culture and Empire interests, Michelle J. Smith, *Empire in British Girls' Literature and Culture: Imperial Girls, 1880–1915* (Houndmills: Palgrave Macmillan, 2011).
64. Baden-Powell and Baden-Powell, *The Handbook for Girl Guides*, pp. 320–7.
65. Smith, *Empire in British Girls' Literature and Culture*, p. 139.
66. Flora Lucy Freeman, *Religious and Social Work amongst Girls* (London: Skeffington & Son, 1901), pp. 45, 47.
67. Baron, *The Growing Generation*, pp. 23, 101.
68. MRC: YWCA, MSS.243/5/4, *Our Own Gazette*, 10 (September 1893), Le Garde, 'Outdoor Sports', 107.
69. Miss Constance Bishop, 'Hockey and Rounders', *Girls' Club News*, No. 12 (February 1913), p. 7.
70. WL, GFS, 5GFS/10/41 (Box 208), *The G.F.S. Magazine*, XLIII: 506 (October 1918), Hazel Grimwood, 'Hockey', 171.
71. MRC: YWCA, MSS.243/4/1/1, *Our Own Gazette*, 38 (September 1920), Z. Swainston, 'How an Association like the Y.W.C.A. can meet the needs of the Average Girl', 20–1, quotes on p. 20.

72. Beatrice Webb, *Health of Working Girls: A Handbook for Welfare Supervisors and Others* (London: Methuen, 1917), p. 101.
73. Freeman, *Religious and Social Work amongst Girls*, p. 65.
74. Foreword to Dorothea Moore, *Terry the Girl-Guide* (London: James Nisbet, 1912). Cited in Smith, *Empire in British Girls' Literature and Culture*, p. 139.
75. Mavis Kitching, interview, February 1987. Judy Giles, *Women, Identity and Private Life in Britain, 1900–50* (Basingstoke: Macmillan, 1995), p. 40.
76. MRC: YWCA, MSS.243/4/3/3, *Our Own Gazette*, 41 (March 1923), Daphne Milman, 'Y.W.C.A. Drill Competitions', 13.
77. MRC: YWCA, MSS.243/2/1/9, *The YWCA. The Movement in 1924. A Review*, p. 2.
78. Lucy M. Moor, *Girls of Yesterday and To-Day: The Romance of the YWCA* (London: S.W. Partridge, c.1913), pp. 94–5.
79. MRC: YWCA, MSS.243/5/11, *Our Own Gazette*, 37 (1900, no issue no.), Anon., 'The Manchester Girls' Institute', 4–6.
80. Tony Jeffs, 'Oft-Referenced – Rarely Read? Report of the Inter-Departmental Committee on Physical Deterioration 1904', in R. Gilchrist, T. Jeffs and J. Spence (eds), *Drawing on the Past: Studies in the History of Community Youth Work* (Leicester: The National Youth Agency, 2006), pp. 43–59, on p. 55
81. Maude Stanley, *Clubs for Working Girls* (London: Grant Richards, 1904), pp. 76, 87.
82. 'The Bainbridge Seaside Holiday Home', *Girls' Club News*, No. 11 (January 1913), p. 1.
83. Claire Langhamer, *Women's Leisure in England, 1920–60* (Manchester: Manchester University Press, 2000), p. 2.
84. Albinia Hobart-Hampden, 'The Working Girl of To-Day', *The Nineteenth Century and After*, 43: 255 (May 1898), 724–30, quote on p. 724.
85. Anon., 'Our Clubs at Home', *Girls' Club News*, No. 37 (March 1915), 1.
86. Deborah Valenze, *The First Industrial Woman* (New York and Oxford: Oxford University Press, 1995), pp. 99–100, 86, 3.
87. B.S.K. Nollys, 'A Factory Girl's Day', *A London Magazine* (August 1897), 437–44, on p. 438.
88. Stanley, *Clubs for Working Girls*, p. viii. A number of club organisers referred into the 1930s to the wildness and destructive behaviour of the girls, but also suggested that they could be tamed and were resilient, vital and eager. See Iris Dove, 'Sisterhood or Surveillance? The Development of Working Girls' Clubs in London 1880–1939', unpublished University of Greenwich PhD thesis, 1996, p. 40.
89. Chesser, 'Women and Girls in the Factory', p. 517.
90. Ibid., p. 518.
91. Ibid.
92. Edith C. Harvey, 'Industrial Notes. What we owe to Women Factory Inspectors', *Girls' Club News*, No. 13 (March 1913), pp. 55–6.
93. MRC: YWCA, MSS.243/45, Social and Legislation Department: Register of Complaints, March 1919-September 1923.
94. 'Our Clubs at Home. Honor Club', *Girls' Club News*, No. 4 (June 1912), 1; Our Clubs at Home. The Chelsea Girls Club', No. 6 (August 1912), 1; 'Our Clubs at Home. St. John-at-Hackney Girls' Club', No. 21 (November 1913), 1.

95. 'N.O.G.C. Notes', *Girls' Club News*, No. 11 (January 1913), 8.
96. R.K. Brown, 'Domestic Education', *Girls' Club News*, No. 13 (March 1913), 6.
97. Anon., 'The Possibilities of Factory Visiting', *Girls' Club Journal*, II: 5 (May 1910), 33–5, quote on p. 35.
98. Dove, 'Sisterhood or Surveillance?', pp. 172–3. See also M.C. Martin, 'Gender, Religion and Recreation: Flora Lucy Freeman and Female Adolescence 1890–1925', in Gilchrist, Jeffs and Spence (eds), *Drawing on the Past*, 61–77, who makes a similar argument and also challenges Dyhouse's assertion that girls' clubs and Guides were necessarily intended to reinforce femininity.
99. Potter, *Boys in Khaki, Girls in Print*, p. 37.
100. Girlguiding UK Archives, *Girl Guides Gazette*, IV: 48 (December 1917), The Chief Commissioner, 'For Offices. The Future and the Present', 191–3, quote on p. 191. For the role of Girl Scouting in America in preparing girls for their future as housewives and mothers, see Rima D. Apple, *Perfect Motherhood: Science and Childrearing in America* (New Brunswick, NJ and London: Rutgers University Press, 2006), pp. 67–71.
101. Baden-Powell and Baden-Powell, *The Handbook for Girl Guides*, p. 320.
102. Smith, *Empire in British Girls' Literature and Culture*, pp. 138–9.
103. Martin, 'Gender, Religion and Recreation', pp. 71, 61, 63.
104. MRC: YWCA MSS.243/2/1/14, *The YWCA in 1929. A Review*, p. 5.
105. Angela Woollacott, *On Her Their Lives Depend. Munitions Workers in the Great War* (Berkeley, CA and London: University of California Press, 1994); Braybon, *Women Workers in the First World War*; Braybon and Summerfield, *Out of the Cage*; J.M. Winter, *The Great War and the British People* (Houndmills: Macmillan, 1985), p. 46.
106. Ministry of Reconstruction, *Report of the Women's Employment Committee*, Cmd. 135 (1919), 80. Cited in Woollacott, *On Her Their Lives Depend*, p. 17.
107. Ineson and Thom, 'T.N.T. Poisoning and the Employment of Women Workers'; Braybon and Summerfield, *Out of the Cage*, p. 80.
108. Dorothy Poole, typescript, Imperial War Museum, Wom. Coll. MUN 17. Cited in Braybon and Summerfield, *Out of the Cage*, pp. 81–2.
109. *Common Cause* (29 September 1916), p. 305. Cited in Woollacott, *On Her Their Lives Depend*, p. 82.
110. Braybon and Summerfield, *Out of the Cage*, p. 82.
111. Angela Woollacott, ' "Khaki Fever" and its Control: Gender, Class, Age and Sexual Morality on the British Homefront in the First World War', *Journal of Contemporary History*, 29 (1994), 325–47.
112. Angela Woollacott, 'Maternalism, Professionalism and Industrial Welfare Supervisors in World War One Britain', *Women's History Review*, 3 (1994), 29–56; Vicky Long, 'Industrial Homes, Domestic Factories: The Convergence of Public and Private Space in Interwar Britain', *Journal of British Studies*, 50 (2011), 434–64, p. 437. These women implemented the suggestions for improvement of the workplace, made by the lady Factory Inspectors appointed after 1893 to improve working conditions for women: Jones, 'Women Health Workers'; Ruth Livesey, 'The Politics of Work: Feminism, Professionalization and Women Inspectors of Factories and Workshops', *Women's History Review*, 13 (2004), 233–61.

113. Braybon and Summerfield, *Out of the Cage*, p. 88. See also Long, *The Rise and Fall of the Healthy Factory*, Chapter 1.
114. J. Kennedy Maclean and T. Wilkinson Riddle, *The Second Picture of War (The Story of the Y.W.C.A. War Service)* (London, Edinburgh and New York: Marshall Brothers, 1919), p. 16.
115. MRC: YWCA MSS.243/2/1/5, *A Review 1914–1920*, p. 13.
116. MRC: YWCA MSS.243/2/1/6, *A Review YWCA 1920–1921*, p. 12.
117. MRC: YWCA MSS.243/2/1/4, *A Review YWCA 1918*, p. 7.
118. MRC: YWCA MSS.243/2/1/3/1, *A Review 1917*, p. 16.
119. MRC: YWCA MSS.243/64/38, Y.W.C.A. War Time Work (promotion leaflet) c.1915.
120. Long, 'Industrial Homes, Domestic Factories', pp. 444–5. See also, for canteens, James Vernon, *Hunger: A Modern History* (Cambridge, MA and London: Harvard University Press, 2007), pp. 161–80.
121. MRC: YWCA MSS.243/2/1/3/1, *A Review 1917*, p. 17.
122. Arthur Marwick, *Women at War*, 1914–1918 (London: Fontana, for the Imperial War Museum, 1977), p. 96.
123. Monica Cosens, *Lloyd George's Munition Girls* (London: Hutchinson, 1916), pp. 58–9.
124. MRC: YWCA MSS.243/2/1/4, *A Review YWCA 1918*, p. 6.
125. MRC: YWCA MSS.243/2/1/4, *A Review YWCA 1918*, p. 26.
126. MRC: YWCA MSS.243/2/1/5, *A Review 1914–1920*, p. 13.
127. Anon., 'Recreations and Enjoyment in War Time', *Girls' Club News*, No. 35, (January 1915), pp. 3–4.
128. Webb, *Health of Working Girls*, foreword.
129. Ibid., p. 99.
130. Ibid., pp. 76–7.
131. D.J. Collier, *The Girl in Industry* (London: G. Bell and Sons, 1918), pp. 8, 24, 30–7, quote on p. 34.
132. Sarah MacDonald, *Simple Health Talks with Women War Workers* (London: Methuen, 1917), pp. 12, 14–5, 30–1.
133. Braybon, *Women Workers in the First World War*, p. 146.
134. MRC: YWCA MSS.243/2/1/4, *A Review YWCA 1918*, p. 19.
135. MRC: YWCA MSS.243/2/1/4, *A Review YWCA 1918*, pp. 26–7.
136. MRC: YWCA MSS.243/2/1/6, *A Review YWCA 1920–1921*, p. 4.
137. MacDonald, *Simple Health Talks*, pp. 73, 54, 61.
138. Mary Scharlieb and Barbara Butts, *England's Girls and England's Future* (London: National Council for Combatting Venereal Diseases, 1917), pp. 6–11, quote on p. 10.
139. Mary Scharlieb, *Venereal Diseases in Children and Adolescents: Their Recognition and Prevention (Notes of Three Lectures addressed to Schoolmistresses, at the Royal Society of Medicine, London, September 1916)* (London: National Council for Combating Venereal Diseases, 1916), p. 15.
140. Mary Scharlieb, *The Challenge of War-Time to Women* (London: National Council of Public Morals for Great and Greater Britain, 1916); Scharlieb and Butts, *England's Girls and England's Future*.
141. Long, 'Industrial Homes, Domestic Factories', pp. 435–6.

142. James Shelley, 'From Home Life to Industrial Life: With Special Reference to the Adolescent Girl', in J.J. Findlay (ed.), *The Young Wage Earner And the Problem of his Education* (London: Sidgwick & Jackson, 1918), p. 23.

143. Helen J. Ferris, *Girls' Clubs: Their Organization and Management. A Manual for Workers*, with an introduction by Jane Deeter Rippin (New York: E.P. Dutton and London: J.M. Dent, 1918), p. 255.

144. Emily Matthias, 'The Young Factory Girl', in Findlay (ed.), *The Young Wage Earner*, pp. 87, 88.

145. Ferris, *Girls' Clubs*, pp. 117–8. Though this volume focused on US organisations of girls clubs, it appears to have been also directed at club leaders in Britain.

146. Matthias, 'The Young Factory Girl', pp. 93, 99.

147. MRC: YWCA, MSS.243/4/1/10, *Our Own Gazette*, 38 (September 1920), Swainston, 'How an Association', p. 20.

148. Ibid., (October 1920), Agnes M. Miall, 'Good Wives, Good Mothers and Good Citizens', 15.

149. Gertrude and Godfrey Pain, *Girls' Clubs. A Practical Handbook for Workers Among Girls of Eleven to Fourteen, Including Games and List of Yarns* (London: Ludgate Circus House, 1932), pp. 10–13.

6 Conclusion: Future Mothers of the Empire or a 'Double Gain'?

1. Florence Harvey Richards, *Hygiene for Girls. Individual and Community* (London: D.C. Health, 1913), p. 9.

2. Bim Andrews, 'Making Do'. Cited in John Burnett (ed.), *Destiny Obscure: Autobiographies of Childhood, Education and Family from the 1820s to the 1920s* (London and New York: Routledge, 1982), pp. 120, 123; Mavis Kitching, interview, February 1987. Judy Giles, *Women, Identity and Private Life in Britain, 1900–50* (Basingstoke: Macmillan, 1995), p. 40.

3. Sally Mitchell, *The New Girl: Girls' Culture in England 1880–1915* (New York: Columbia University Press, 1995); Michelle Smith, *Empire in British Girls' Literature and Culture: Imperial Girls, 1880–1915* (Houndmills: Palgrave Macmillan, 2011). See also for the renaissance of motherhood during the period 1870–1920, Christina Hardyment, *Dream Babies: Child Care from Locke to Spock* (London: Jonathan Cape, 1983).

4. M. Scharlieb, 'Recreational Activities of Girls during Adolescence', *Child-Study: The Journal of the Child-Study Society*, 4 (1911), 1–14, quotes on pp. 8, 9.

5. Mrs Eric Pritchard, 'Physical Culture for Women', *Pall Mall Magazine*, 33 (May 1904), 141–4, quote on p. 142.

6. Annie Burns Smith, *Talks with Girls upon Personal Hygiene* (London: Pitman, 1912) (with a Preface by John Robertson, MD, Medical Officer of Health, City of Birmingham, and an Introduction by Mary D. Sturge, Physician to the Birmingham and Midland Hospital for Women), pp. 2, 5.

7. Such incursions have been illuminated by Jane Lewis and Deborah Dwork, for example, in the British context and Rima Apple for the US: Jane Lewis,

The Politics of Motherhood: Child and Maternal Welfare in England 1900–1939 (London: Croom Helm, 1980); Deborah Dwork, *War is Good for Babies and Other Young Children: A History of the Infant and Child Welfare Movement in England 1898–1918* (London and New York: Tavistock, 1987); Rima D. Apple, *Mothers and Medicine: A Social History of Infant Feeding 1890–1950* (Madison, WI: University of Wisconsin Press, 1987).

8. Agnes L. Stenhouse and E. Stenhouse, *A Health Reader for Girls* (London: Macmillan, 1918), pp. 3, 5.

9. Ibid., p. 2.

10. Ibid., pp. 3–4.

11. Ibid., p. 4.

12. Mary Humphreys, *Personal Hygiene for Girls* (London, New York, Toronto and Melbourne: Cassell, 1913), p. 1 (her emphasis).

13. Ibid., preface.

14. Ibid., p. 54.

15. Ibid., p. 137.

16. Mary Scharlieb and F. Arthur Sibly, *Youth and Sex: Dangers and Safeguards for Boys and Girls* (London: T.C. & E.C. Jack, 1919), p. 26.

17. Elizabeth Sloan Chesser, *Physiology and Hygiene for Girls' Schools and Colleges* (London: G. Bell and Sons, 1914), p. 87.

18. Ibid.

19. Greta Jones, 'Women and Eugenics in Britain: The Case of Mary Scharlieb, Elizabeth Sloan Chesser, and Stella Browne', *Annals of Science*, 51 (1995), 481–502, p. 489.

20. Elizabeth Sloan Chesser, *From Girlhood to Womanhood* (London, New York, Toronto and Melbourne: Cassell, 1913), pp. 79–80.

21. Christine M. Murrell, *Womanhood and Health* (London: Mills & Boon, 1923).

22. V.A. Zelizer, *Pricing the Priceless Child: The Changing Social Value of Children* (New York: Basic Books, 1985); Carolyn Steedman, 'Bodies, Figures and Physiology: Margaret McMillan and the Late Nineteenth-Century Remaking of Working-Class Childhood', in Roger Cooter (ed.), *In the Name of the Child: Health and Welfare 1880–1940* (London and New York: Routledge, 1992), pp. 19–44.

Bibliography

Archives

Modern Records Centre, University of Warwick
Association of Assistant Mistresses (MSS.59)
Association of Headmistresses (MSS.188)
Association of Teachers of Domestic Science (MS.177)
National Cycle Archive (MSS.328)
Young Women's Christian Association (Platform 51) (MSS.243)

Wellcome Library
Archives of the Medical Women's Federation (MSS.SA/MWF)

Girlguiding UK Archives

The Women's Library
The Girls' Friendly Society (GFS)

North London Collegiate School Archives (NLCSA)
Official papers

1887 [C.5158] Elementary Education Acts. Third Report of the Royal Commission Appointed to Inquire into the Working of the Elementary Education Acts, England and Wales.
Mrs Ely-Dallas, Organising Teacher of Physical Exercises under the School Board for London, 'Physical Education in the Girls' and Infants' Departments in the Schools of the London School Board', in *Special Reports on Educational Subjects*, vol. 2 (London: Her Majesty's Stationary Office, 1898).
1906 [Cd. 3036] Factories and Workshops. Annual Report of the Chief Inspector of Factories and Workshops for the Year 1905. Reports and Statistics.

Periodicals and Newspapers

Blackwood's Magazine
British Medical Journal
Chambers's Journal
Cyclists' Touring Club Gazette
Cyclists' Touring Club Monthly Gazette and Official Record
Girls' Club News
Girl's Own Paper
Girl's Realm
Good Health

Health
Lady's Realm
Lancet
London Journal
Popular Science Monthly
Pall Mall Magazine
Queen
Review of Reviews
The Badminton Magazine of Sports and Pastimes
The Contemporary Review
The Domestic Magazine and Journal for the Household
The Hub
The Idler
The Leisure Hour
The Nineteenth Century
The Nineteenth Century and After
The Speaker
The Times
The Woman's Signal
Westminster Review

Contemporary books and articles

Alexander, A., *Healthful Exercises for Girls* (London: George Philip & Son, 1886).

Anderson, Elizabeth Garrett, 'Sex in Mind and Education: A Reply', *Fortnightly Review*, new series, 15 (May 1874), 582–94.

Anon. [E. Lynn Linton], 'The Girl of the Period', *Saturday Review*, 25 (14 March 1868), 339–40.

Anon., *The Ladies' Physician. A Guide for Women in the Treatment of their Ailments. By a London Physician*, 9th edn (London, Paris & Melbourne: Cassell, 1891).

A Specialist [Anon.], *Beauty and Hygiene for Women and Girls* (London: Swan Sonnenschein & Co., 1893).

Anon., 'Are Athletics Over-done in Girls' Colleges and Schools? A Symposium', *Woman at Home*, 6 (April 1912), 247–52.

Baden-Powell, Agnes Smyth, and Sir Robert Baden-Powell, *The Handbook for Girl Guides; or How Girls Can Help Build the Empire* (London: Thomas Nelson and Sons, 1912).

Barnard, Amy B., *The Girl's Encyclopaedia* (London: Pilgrim, 1909).

Barnard, A.B., *The Girl's Book About Herself* (London: Cassell, 1912).

Baron, Barclay, *The Growing Generation: A Study of Working Boys and Girls in our Cities* (London: Student Christian Movement, 1911).

Bathhurst, K., 'The Physique of Girls', *Nineteenth Century and After* (May 1906), 825–33.

Beale, D., *On the Education of Girls: A Paper Read at the Social Science Congress, October, 1865, and reprinted from the Transactions* (London: Bell & Daldy, 1866).

Beale, Dorothea, *Address to Parents* (London: George Bell & Sons, 1888).

Black, George, *Household Medicine: A Guide to Good Health, Long Life, and the Proper Treatment of all Diseases and Accidents* (London: Ward, Lock & Co., 1883).

Black, George, *The Young Wife's Advice Book: A Guide for Mothers on Health and Self-Management*, 6th edn (London: Ward, Lock and Co., 1888).

Black, George, *The Young Wife's Advice Book. A Complete Guide for Mothers on Health, Self-Management, and the Care of the Baby* (London, Melbourne and Toronto: Ward, Lock and Co., 1910).

Blackwell, Elizabeth, *Lectures on the Laws of Life With Special Reference to the Physical Education of Girls* (London: Sampson, Low, Son, & Marsten, 1871).

Blanchard, Phyllis, *The Care of the Adolescent Girl: A Book for Teachers, Parents, and Guardians*, with prefaces by Mary Scharlieb and G. Stanley Hall (London: K. Paul, Trench, Trübner, 1921).

Bourne, E.D. (ed.), *Girls' Games: A Recreation Handbook for Teachers and Scholars* (London: Griffith, Farran, Okeden & Welsh, 1887).

Bryant, S., *Over-work: From the Teacher's Point of View with Special Reference to the Work in Schools for Girls, A Lecture Delivered at the College of Preceptors, November 19th, 1884* (London: Francis Hodgson, 1885).

Burstall, Sara, *English High Schools for Girls: Their Aims, Organisation, and Management* (London, New York, Bombay, and Calcutta: Longmans, Green, and Co., 1907).

Cantlie, James, 'The Influence of Exercise on Health', in Malcolm Morris (ed.), *The Book of Health* (London, Paris and New York: Cassell, 1883), pp. 381–461.

Caulfeild, S.F.A., *A Directory of Girls' Societies, Clubs, and Unions, Conducted on Unprofessional Principles* (London: Griffith, Farran, Okeden & Welsh, 1886).

Chant, L. Ormiston, 'Woman as an Athlete: A Reply to Dr. Arabella Kenealy', *The Nineteenth Century* (May 1899), 745–54.

Chavasse, Pye Henry, *Advice to a Mother on the Management of Her Children and on the Treatment on the Moment of Some of their More Pressing Illnesses and Accidents*, 14th edn (London: J. & A. Churchill, 1889).

Chavasse, Pye Henry, *Advice to a Mother: On the Management of Her Children, and on the Treatment on the Moment of some of their more Pressing Illnesses and Accidents*, 9th edn (London: Cassell, 1911).

Chesser, Elizabeth Sloan, 'Women and Girls in the Factory', *Westminster Review*, 173: 5 (May 1910), 516–19.

Chesser, Elizabeth Sloan, *Perfect Health for Women and Children* (London: Methuen, 1912).

Chesser, Elizabeth Sloan, *From Girlhood to Womanhood* (London, New York, Toronto and Melbourne: Cassell, 1913).

Chesser, Elizabeth Sloan, *Woman, Marriage and Motherhood* (London, New York, Toronto and Melbourne: Cassell, 1913).

Chesser, Elizabeth Sloan, *Physiology and Hygiene for Girls' Schools and Colleges* (London: G. Bell and Sons, 1914).

Chesser, Elizabeth Sloan, *Vitality: A Book on the Health of Women and Children* (London: Methuen, 1935).

Chisholm, Catherine, *The Medical Inspection of Girls in Secondary Schools* (London, New York, Bombay and Calcutta: Longmans, Green, and Co., 1914).

Chreiman, Miss, *Physical Education of Girls. A Lecture Delivered in the Lecture Room of the Exhibition, July 25th, 1884*, Published for the Executive Council of the International Health Exhibition and the Council of the Society of Arts (London: William Clowes and Sons, 1884).

Chreiman, Miss, *Physical Culture of Women: Lecture, by Request of the Council of Parkes' Museum, March 15th, 1888* (London: Sampson Low, Marston, Searle, & Rivington, 1888).

Clark, Andrew, 'Anaemia or Chlorosis of Girls, Occurring more Commonly between the Advent of Menstruation and the Consummation of Womanhood', *Lancet*, 2 (19 November 1887), 1003–5.

Clarke, E., *Sex in Education: Or, a Fair Chance for Girls* (Boston, MA: James R. Osgood, 1874).

Clouston, T.S., 'Puberty and Adolescence Medico-Psychologically Considered', *Edinburgh Medical Journal*, 26: 1 (July 1880), 5–17.

Clouston, T.S., 'Female Education from a Medical Point of View', *Popular Science Monthly*, 24 (December 1883, January 1884), 214–28, 319–34.

Clouston, T.S., *Clinical Lectures on Mental Diseases*, 3rd edn (London: J. and A. Churchill, 1892).

Collier, D.J., *The Girl in Industry* (London: G. Bell and Sons, 1918).

Corner, Edred M., 'Physical Exercises' in T.C. Allbutt and H.D. Rolleston (eds), *A System of Medicine by Many Writers*, vol. 1 (London and New York: Macmillan, 1905), pp. 382–421.

Cosens, Monica, *Lloyd George's Munition Girls* (London: Hutchinson, 1916).

Countess of Malmesbury, Susan, G. Lacy Hillier, and H. Graves, *Cycling* (London: Lawrence and Bullen, 1898).

Crichton-Browne, J., 'Education and the Nervous System', in Malcolm Morris (ed.), *The Book of Health* (London, Paris and New York: Cassell, 1883), pp. 269–380.

Crichton-Browne, Sir James, 'An Oration on Sex in Education. Delivered at the Medical Society of London on May 2nd, 1892', *Lancet*, 1 (7 May 1892), 1011–18.

Curwen, Maud and Ethel Herbert, *Simple Health Rules and Health Exercises for Busy Women and Girls* (London: Simpkin, Marshall, Hamilton, Kent & Co., 1912).

Dickinson, Robert, 'Bicycling for Women from the Standpoint of the Gynecologist', *American Journal of Obstetrics*, 31 (1895), 24–35.

Donkin, H.B., 'Hysteria', in D. Hack Tuke (ed.), *A Dictionary of Psychological Medicine* (London: J. & A. Churchill, 1892), pp. 618–27.

Dove, Jane Francis, 'Cultivation of the Body', in Dorothea Beale, Lucy H.M. Soulsby and Jane Francis Dove, *Work and Play in Girls' Schools by Three Head Mistresses* (London, New York and Bombay: Longmans, Green, and Co., 1898), pp. 396–423.

Duffey, E.B., *What Women Should Know. A Woman's Book about Women. Containing Practical Information for Wives and Mothers* (Philadelphia, PA: J.M. Stoddart [1873]).

Dukes, Clement, 'Health at School', in Malcolm Morris (ed.), *The Book of Health* (London, Paris and New York: Cassell, 1883), pp. 677–725.

Dukes, Clement, *Health at School Considered in its Mental, Moral, and Physical Aspects* (London, Paris, New York & Melbourne: Cassell, 1887).

Dukes, Clement, *Work and Overwork in Relation to Health in Schools*, An Address Delivered before the Teachers' Guild of Great Britain and Ireland at its Fifth General Conference Held in Oxford, 17–20 April 1893 (London: Percival, 1893).

Dukes, Clement, *The Essentials of School Diet or the Diet Suitable for the Growth and Development of Youth*, 2nd edn (London: Rivingtons, 1899).

Dukes, Clement, 'The Hygiene of Youth', in T.C. Allbutt and H.D. Rolleston (eds), *A System of Medicine by Many Writers*, vol. 1 (London and New York: Macmillan, 1905), pp. 161–81.

Farningham, Marianne, *Girlhood* (London: James Clarke, 1869).

Farningham, Marianne, *Girlhood* (London: James Clarke, 1895).

Fenton, W.H., 'A Medical View of Cycling for Ladies', *The Nineteenth Century* (January–June 1896), 796–801.

Ferris, Helen J., *Girls' Clubs: Their Organization and Management. A Manual for Workers*, with an introduction by Jane Deeter Rippin (New York: E.P. Dutton and London: J.M. Dent, 1918).

Findlay, J.J. (ed.), *The Young Wage Earner And the Problem of his Education* (London: Sidgwick & Jackson, 1918).

Findlay, M.E., 'The Education of Girls', *The Paidologist*, VII (1905), 83–93.

Fletcher, J. Hamilton, 'Feminine Athletics', *Good Words*, 20 (January 1879), 533–6.

Freeman, Flora Lucy, *Religious and Social Work amongst Girls* (London: Skeffington & Son, 1901).

Galabin, A.L., *Diseases of Women*, 5th edn (London: J. & A. Churchill, 1893).

Galbraith, Anna M., *Hygiene and Physical Culture for Women* (London: B.F. Stevens, 1895).

Grand, Sarah, *The Heavenly Twins* (London: George Heinemann, 1893).

Gull, William, 'Anorexia Nervosa (Apepsia Hysterica, Anorexia Hysteria)', *Transactions of the Clinical Society of London*, 7 (1874), 22–8.

Gull, William, 'Anorexia Nervosa', *Lancet*, 1 (17 March 1888), 516–17.

Hall, G. Stanley, *Adolescence: Its Psychology and the Relation to Physiology, Anthropology, Sociology, Sex, Crime, Religion and Education* (New York: D. Appleton, 1904).

Hall, G. Stanley, *Youth: Its Education, Regimen, and Hygiene* (New York: D. Appleton, 1906).

Hall, G. Stanley, *Educational Problems*, vol. II (New York and London: D. Appleton, 1911).

Hallam, Margaret, *Health and Beauty for Women and Girls: A Course of Physical Culture* (London: C. Arthur Pearson, 1921).

Hallam, Margaret, *Dear Daughter of Eve: A Compleat Book of Health and Beauty* (London: W. Collins, 1924).

Herrick, Christine Terhune, 'Women in Athletics', *The Idler* (October 1902), 677–80.

Hewitt Griffin, H., *Cycles and Cycling. With a Chapter for Ladies*, by Miss L.C. Davidson (London: George Bell & Sons, 1890).

Hoggan, Francis Elizabeth, *On the Physical Education of Girls, Read at the Annual Meeting of the Fröbel Society*, on December 9, 1879 (London: W. Swan Sonnenschein & Allen, 1880).

Humphreys, Mary, *Personal Hygiene for Girls* (London, New York, Toronto and Melbourne: Cassell, 1913).

Jackson, A. Reeves, 'Diseases Peculiar to Women', in Frederick A. Castle (ed.), *Wood's Household Practice of Medicine Hygiene and Surgery: A Practical Treatise for the Use of Families, Travellers, Seamen, Miners and Others*, vol. 2 (London: Sampson Low, Marsten, Searle, & Rivington, 1881), pp. 538–43.

Johnson, Edward, *The Hydropathic Treatment of Diseases Peculiar to Women; and of Women in Childbed; with some Observations on the Management of Infants* (London: Simpkin, Marshall, and Co., 1850).

Johnson, Walter, *The Morbid Emotions of Women; Their Origin, Tendencies, and Treatment* (London: Simpkin, Marshall, and Co., 1850).

Jones, Robert, 'Girls' Schools, Games, and Neurasthenia', *Lancet*, 1 (4 February 1911), 329.

Kellerman, Annette, *Physical Beauty and How to Keep It* (London: Heinemann, 1918).

Kelly, Howard A., *Medical Gynecology* (New York and London: Appleton & Co., 1908).

Kelynack, T.N., *Youth* (London: Charles H. Kelly, 1918).

Kenealy, Arabella, 'Woman as Athlete', *The Nineteenth Century* (April 1899), 635–45.

Kenealy, Arabella, 'Woman as an Athlete: A Rejoinder', *The Nineteenth Century* (June 1899), 915–29.

Kenealy, Arabella, *Feminism and Sex-Extinction* (London: T. Fisher Unwin [1920]).

Kingsford, Anna Bonus, *Health, Beauty and the Toilet: Letters to Ladies from a Lady Doctor* (London and New York: Frederick Warne, 1886).

Kirk, Edward Bruce, *A Talk with Girls about Themselves* (London: Simpkin, Marshall, Hamilton, Kent & Cov., 1895).

Linton, E. Lynn, *The Girl of the Period and Other Social Essays*, vol. 1 (London: Richard Bentley & Son, 1883).

Lowe, Ernest W., 'Women and Exercise', *Chambers's Journal*, 3: 108 (23 December 1899), 49–52.

MacDonald, Sarah, *Simple Health Talks with Women War Workers* (London: Methuen, 1917).

MacFadden, Bernarr and Marion Malcolm, *Health – Beauty – Sexuality: From Girlhood to Womanhood* (New York: Physical Culture Publishing Company, 1904).

Maclean, J. Kennedy and Riddle, T. Wilkinson, *The Second Picture of War (The Story of the Y.W.C.A. War Service)* (London, Edinburgh and New York: Marshall Brothers, 1919).

MacNaughton-Jones, H., 'The Relation of Puberty and the Menopause to Neurasthenia', *Lancet*, 1 (29 March 1913), 879–81.

Maudsley, Henry, *The Physiology and Pathology of Mind: A Study of its Distempers, Deformities and Disorders*, 2nd edn (London: Macmillan, 1868).

Maudsley, Henry, 'Sex in Mind and in Education', *Fortnightly Review*, new series, 15 (April 1874), 466–83.

Maudsley, Henry, *The Pathology of Mind: A Study of its Distempers, Deformities and Disorders*, 3rd edn of 2nd part of *The Physiology and Pathology of Mind* (London: Macmillan 1879).

Moor, Lucy M., *Girls of Yesterday and To-Day: The Romance of the YWCA* (London: S.W. Partridge, c.1913).

Morris, Malcolm (ed.), *The Book of Health* (London, Paris and New York: Cassell, 1883).

Mortimer-Granville, J., *Youth: Its Care and Culture. An Outline of Principles for Parents and Guardians* (London: David Bogue, 1880).

Murrell, Christine M., *Womanhood and Health* (London: Mills & Boon, 1923).

Olsen, Alfred B. and M. Ellsworth Olsen (eds), *The School of Health: A Guide to Health in the Home* (London: International Tract Society, 1906).

Pain, Gertrude and Godfrey, *Girls' Clubs. A Practical Handbook for Workers Among Girls of Eleven to Fourteen, Including Games and List of Yarns* (London: Ludgate Circus House, 1932).

Peters, Charles (ed.), *The Girl's Own Outdoor Book Containing Practical Help to Girls on Matters relating to Outdoor Occupation and Recreation* (London: The Religious Tract Society, 1889).

Pfeiffer, Emily, *Women and Work: An Essay Treating on the Relation to Health and Physical Development, of the Higher Education of Girls, and the Intellectual or More Systematised Effort of Women* (London: Trübner & Co., 1888).

Playfair, W.S., *The Systematic Treatment of Nerve Prostration and Hysteria* (London: Smith Elder, & Co., 1883).

Playfair, W.S., 'Note on the So-Called "Anorexia Nervosa"', *Lancet*, 1 (28 April 1888), 817–18.

Playfair, W.S., 'The Systematic Treatment of Functional Neurosis', in D.H. Tuke (ed.), *A Dictionary of Psychological Medicine*, vol. II (London: J. & A. Churchill, 1892), pp. 850–7.

Playfair, W.S., 'Remarks on the Education and Training of Girls of the Easy Classes at and about the Period of Puberty', *British Medical Journal*, 2 (7 December 1895), 1408–10.

Playfair, W.S., 'The Nervous System in Relation to Gynaecology', in Thomas Clifford Allbutt and W.S. Playfair (eds), *A System of Gynaecology by Many Writers* (London: Macmillan, 1896), pp. 220–31.

Pritchard, Mrs Eric, 'Physical Culture for Women', *Pall Mall Magazine*, 33 (May 1904), 141–4.

Rentoul, Robert Reid, *The Dignity of Woman's Health and the Nemesis of its Neglect (A Pamphlet for Women and Girls)* (London: J. & A. Churchill, 1890).

Rentoul, Robert Reid, *Race Culture; Or, Race Suicide? (A Plea for the Unborn)* (London and Felling-on-Tyne: Walter Scott Publishing, 1906).

'Reports issued by the Schools' Inquiry Commission, on the Education of Girls. With Extracts from the Evidence and a Preface (Taunton Report)', by D. Beale, Principle of the Ladies' College Cheltenham (London: David Nutt, 1869).

Richards, Florence Harvey, *Hygiene for Girls. Individual and Community* (London: D.C. Health, 1913).

Ridley, Anne E., *Frances Mary Buss and her Work for Education* (London: Longmans, Green and Co, 1895).

Ruddock, E.H., *The Lady's Manual of Homoeopathic Treatment in the Various Derangements Incident to her Sex*, 9th edn (London: The Homoeopathic Publishing Company, 1886).

Scharlieb, Mary, 'Adolescent Girlhood under Modern Conditions, with Special Reference to Motherhood', *Eugenics Review*, 1 (1909), 174–83.

Scharlieb, M., 'Recreational Activities of Girls during Adolescence', *Child-Study: The Journal of the Child-Study Society*, 4 (1911), 1–14.

Scharlieb, Mary A.D., 'Adolescent Girls from the View-Point of the Physician', *The Child*, 1: 12 (September 1911), 1013–31.

Scharlieb, Mary, *The Challenge of War-Time to Women* (London: National Council of Public Morals for Great and Greater Britain, 1916).

Scharlieb, Mary, *Venereal Diseases in Children and Adolescents: Their Recognition and Prevention (Notes of Three Lectures addressed to Schoolmistresses at the Royal Society of Medicine, London, September, 1916)* (London: National Council for Combating Venereal Diseases, 1916).

Scharlieb, Mary and Barbara Butts, *England's Girls and England's Future* (London: National Council for Combatting Venereal Diseases, 1917).

Scharlieb, Mary and F. Arthur Sibly, *Youth and Sex: Dangers and Safeguards for Boys and Girls* (London: T.C. & E.C. Jack, 1919).

Schofield, A.T., 'Nervousness and Hysteria', *The Leisure Hour* (June 1889), 412–15.

Schofield, Alfred T., 'The Modern Development of Athletics', *The Leisure Hour* (October 1891), 817–20.

Shadwell, A. 'The Hidden Dangers of Cycling', *The National Review*, 28: 168 (February 1897), 787–96.

Skey, F.C., *Hysteria. Six Lectures Delivered to the Students of St. Bartholomew's Hospital, 1866* (London: Longmans, Green, Reader & Dyer, 1867).

Slaughter, J.W., *The Adolescent* (London: G. Allen, 1907).

Smith, Annie Burns, *Talks with Girls upon Personal Hygiene* (London: Pitman, 1912).

Spencer, Herbert, *Education: Intellectual, Moral and Physical* (London: G. Manwaring, 1861).

Spicer, Howard (ed.), *Sports for Girls*, with an introduction by Mrs Ada S. Ballin (London: Andrew Melrose, 1900).

Stables, Gordon, *Rota Vitae: The Cyclists Guide to Health and Rational Enjoyment* (London: Iliffe & Son, 1886).

Stables, Gordon, *Health upon Wheels; or, Cycling a Means of Maintaining the Health* (London: Iliffe & Son, 1887).

Stables, Gordon, *The People's A B C Guide to Health* (London: Hodder and Stoughton, 1887).

Stables, Gordon, *The Girl's Own Book of Health and Beauty* (London: Jarrold and Sons, 1891).

Stables, Gordon, *The Mother's Book of Health and Family Advisor* (London: Jarrold and Sons, 1894).

Stables, Gordon, *The Wife's Guide to Health and Happiness* (London: Jarrold and Sons, 1894).

Stables, Gordon, *Heartstone Talks on Health and Home* (Norwich: Jarrold and Sons, 1904).

Stanley, Maude, *Clubs for Working Girls* (London: Grant Richards, 1904).

Stenhouse, Agnes L. and E. Stenhouse, *A Health Reader for Girls* (London: Macmillan, 1918).

Sturge, V. 'The Physical Education of Women' [n.d.], in Dale Spender (ed.), *The Education Papers: Women's Quest for Equality in Britain, 1850–1912* (New York and London: Routledge & Kegan Paul, 1987), pp. 284–94.

Tait, Robert Lawson, *The Diseases of Women*, 2nd edn (New York: William Wood & Co., 1879).

Thackrah, Margaret G., 'The Danger of Athletics for Girls', *Lancet*, 1 (18 June 1921), 1328.

The Girl's Realm Annual (November 1898–October 1899) (London: Hutchinson, 1899).

The Girl's Realm Annual (November 1899–October 1990) (London: S.H. Bousfield, 1900).

Thorburn, John, *Female Education from a Medical Point of View* (Manchester: J.E. Cornish, 1884).

Thorburn, John, *Female Education in its Physiological Aspect* (Manchester: J.E. Cornish, 1884).

Thorburn, John, *A Practical Treatise on the Diseases of Women* (London: Charles Griffin, 1885).

Tilt, E.J., *On the Preservation of the Health of Women at the Critical Periods of Life* (London: John Churchill, 1851).

Tilt, E.J., *Elements of Health, and Principles of Female Hygiene* (London: Henry G. Bohn, 1852).

Tilt, Edward John, 'Education of Girls', *British Medical Journal*, 1 (2 April 1887), 751.

Treves, Frederick, 'The Influence of Dress on Health', in Malcolm Morris (ed.), *The Book of Health* (London, Paris and New York: Cassell, 1883), pp, 461–517.

Walker, Emma E., *Beauty through Hygiene: Common-Sense Ways to Health for Girls* (London: Hutchinson & Co., 1905).

Walsh, J.H., *Domestic Medicine and Surgery: With a Glossary of the Terms Used Therein* (London: Frederick Warne, 1866).

Weatherly, Lionel, *The Young Wife's Own Book: A Manual of Personal and Family Hygiene* (London: Griffith and Farran, 1882).

Webb, Beatrice, *Health of Working Girls: A Handbook for Welfare Supervisors and Others* (London: Methuen, 1917).

Whitley, Mary, *Every Girl's Book of Sport, Occupation and Pastime* (London: George Routledge and Sons, 1897).

Select secondary literature

Apple, Rima D., *Mothers and Medicine: A Social History of Infant Feeding 1890–1950* (Madison, WI: University of Wisconsin Press, 1987).

Apple, Rima D., *Perfect Motherhood: Science and Childrearing in America* (New Brunswick, NJ and London: Rutgers University Press, 2006).

Atkinson, Paul, 'Fitness, Feminism and Schooling', in Sara Delamont and Lorna Duffin (eds), *The Nineteenth-Century Woman: Her Cultural and Physical World* (London: Croom Helm, 1978), pp. 92–133.

Bailin, Miriam, *The Sickroom in Victorian Fiction: The Art of Being Ill* (New York: Cambridge University Press, 1994).

Bashford, Alison, *Purity and Pollution: Gender, Embodiment and Victorian Medicine* (Houndmills: Macmillan, 1998).

Beetham, Margaret, *A Magazine of her Own? Domesticity and Desire in the Woman's Magazine, 1800–1914* (London and New York: Routledge, 1996).

Beier, Lucinda McCray, *For Their Own Good: The Transformation of English Working-Class Health Culture, 1880–1970* (Columbus, OH: Ohio State University Press, 2008).

Bevington, Merle Mowbray, *The Saturday Review 1855–1868: Representative Educated Opinion in Victorian England* (New York: AMS Press, 1966).

Bingham, Adrian, *Gender, Modernity, and the Popular Press in Inter-War Britain* (Oxford: Clarendon, 2004).

Bland, Lucy and Lesley A. Hall, 'Eugenics in Britain: The View from the Metropole', in Alison Bashford and Phillipa Levine (eds), *The Oxford Handbook of the History of Eugenics* (Oxford: Oxford University Press, 2010), pp. 213–27.

Bonnell, Marilyn, 'The Power of the Pedal: The Bicycle and the Turn-of-the-Century Woman', *Nineteenth-Century Contexts: An Interdisciplinary Journal*, 14 (1990), 215–39.

Branca, Patricia, *Silent Sisterhood: Middle-Class Women in the Victorian Home* (London: Croom Helm, 1975).

Braybon, Gail, *Women Workers in the First World War* (London: Croom Helm, 1981).

Braybon, Gail and Penny Summerfield, *Out of the Cage: Women's Experiences in Two World Wars* (London and New York: Pandora, 1987).

Bruche, Hilde, *Eating Disorders: Obesity, Anorexia Nervosa, and the Person Within* (New York: Basic Books, 1973).

Brumberg, Joan Jacobs, 'Chlorotic Girls, 1870–1920: A Historical Perspective on Female Adolescence', in Judith Walzer Leavitt (ed.), *Women and Health in America* (Madison, WI: University of Wisconsin Press, 1984), pp. 186–95.

Brumberg, Joan Jacobs, *Fasting Girls: The History of Anorexia Nervosa* (Cambridge, MA: Harvard University Press, 1988).

Brumberg, Joan Jacobs, *The Body Project: An Intimate History of American Girls* (New York: Vintage, 1998).

Brumberg, Joan Jacobs, ' "Something Happens to Girls": Menarche and the Emergence of the Modern Hygienic Imperative', in Judy Walzer Leavitt (ed.), *Women and Health in America*, 2nd edn (Madison, WI: University of Wisconsin Press, 1999), pp. 150–71.

Buckley, Cheryl and Hilary Fawcett, *Fashioning the Feminine: Representation and Women's Fashion from the Fin de Siècle to the Present* (London: I.B. Taurus, 2002).

Burnett, John (ed.), *Destiny Obscure: Autobiographies of Childhood, Education and Family from the 1820s to the 1920s* (London and New York: Routledge, 1982).

Burstyn, Joan N., 'Education and Sex: The Medical Case against Higher Education for Women in England, 1870–1900', *Proceedings of the American Philosophical Society*, 117 (1973), 79–89.

Burstyn, Joan, *Victorian Education and the Ideal of Womenhood* (London: Croom Helm, 1980).

Cadogan, Mary and Patricia Craig, *You're a Brick Angela! The Girls' Story 1839–1985* (London: Victor Gollancz, 1986).

Corke, Helen, *In Our Infancy. An Autobiography, Part I: 1882–1912* (Cambridge: Cambridge University Press, 1975).

Cunningham, Gail, *The New Woman and the Victorian Novel* (London and Basingstoke: Macmillan, 1978).

Davin, Anna, 'Imperialism and Motherhood', *History Workshop Journal*, 5 (1978), 9–66.

Davin, Anna, *Growing Up Poor: Home, School and Street in London 1870–1914* (London: Rivers Oram Press, 1996).

Digby, Anne, 'Women's Biological Straitjacket', in Susan Mendus and Jane Rendall (eds), *Sexuality and Subordination: Interdisciplinary Studies of Gender in the Nineteenth Century* (London and New York: Routledge, 1989), pp. 192–220.

Digby, Anne, *Making a Medical Living: Doctors and Patients in the English Market for Medicine, 1720–1911* (Cambridge: Cambridge University Press, 1994).

Doughty, Terri, *Selections from The Girl's Own Paper, 1880–1907* (Peterborough, ON: Broadview, 2004).

Drotner, Kirsten, *English Children and Their Magazines, 1751–1945* (New Haven, CT and London: Yale University Press, 1988).

Duffin, Lorna, 'The Conspicuous Consumptive: Woman as an Invalid', in Sara Delamont and Lorna Duffin (eds), *The Nineteenth-Century Woman: Her Cultural and Physical World* (London: Croom Helm, 1978), pp. 26–55.

Dwork, Deborah, *War is Good for Babies and Other Young Children: A History of the Infant and Child Welfare Movement in England 1898–1918* (London and New York: Tavistock, 1987).

Dyhouse, Carol, 'Good Wives and Little Mothers: Social Anxieties and the Schoolgirl's Curriculum, 1890–1920', *Oxford Review of Education*, 3 (1977), 21–35.

Dyhouse, Carol, *Girls Growing Up in Late Victorian and Edwardian England* (London: Routledge & Kegan Paul, 1981).

Dyhouse, Carol, *No Distinction of Sex? Women in British Universities, 1870–1939* (London: UCL Press, 1995).

Elston, Mary Ann, ' "Run by Women, (mainly) for Women": Medical Women's Hospitals in Britain, 1866–1948', in Anne Hardy and Lawrence Conrad (eds), *Women and Modern Medicine* (Amsterdam and New York: Rodopi, 2001), pp. 73–107.

Figlio, Karl, 'Chlorosis and Chronic Disease in Nineteenth-Century Britain: The Social Construction of Somatic Illness in a Capitalist Society', *Social History*, 3 (1978), 167–97.

Fildes, Valerie, Lara Marks and Hilary Marland (eds), *Women and Children First: International Maternal and Infant Welfare 1870–1945* (London and New York: Routledge, 1992).

Fletcher, Sheila, *Women First: The Female Tradition in English Physical Education 1800–1980* (London: Athlone, 1984).

Flint, Kate, *The Woman Reader 1837–1914* (Oxford: Clarendon, 1993).

Forrester, Wendy, *Great-Grandmama's Weekly: A Celebration of The Girl's Own Paper 1880–1901* (Guildford and London: Lutterworth Press, 1980).

Foucault, Michel, *The History of Sexuality: Volume 1, An Introduction* (New York: Vintage Books, 1980).

Fowler, David, *The First Teenagers: The Lifestyle of Young Wage-Earners in Interwar Britain* (London: Woburn Press, 1995).

Fowler, David, *Youth Culture in Modern Britain, c.1920–c.1970* (Houndmills: Palgrave Macmillan, 2008).

Frawley, Maria H., *Invalidism and Identity in Nineteenth-Century Britain* (Chicago and London: University of Chicago Press, 2004).

Gijswijt-Hofstra, Marijke and Roy Porter (eds), *Cultures of Neurasthenia From Beard to the First World War* (Amsterdam and New York: Rodopi, 2001).

Gijswift-Hofstra, Marijke and Hilary Marland (eds), *Cultures of Child Health in Britain and the Netherlands in the Twentieth Century* (Amsterdam and New York: Rodopi, 2003).

Giles, Judy, *Women, Identity and Private Life in Britain, 1900–50* (Basingstoke: Macmillan, 1995).

Gillis, John, *Youth and History: Tradition and Change in European Age Relations 1770–Present* (New York and London: Academic Press, 1974).

Gilman, Sander, Helen King, Roy Porter, George S. Rousseau and Elaine Showalter, *Hysteria Beyond Freud* (Berkeley, CA: University of California Press, 1993).

Gomersall, Meg, *Working-Class Girls in Nineteenth-Century England: Life, Work and Schooling* (Houndmills: Macmillan, 1997).

Gorham, Deborah, *The Victorian Girl and the Feminine Ideal* (London and Canberra: Croom Helm, 1982).

Gurjeva, Lyubov G., 'Child Health, Commerce and Family Values: The Domestic Production of the Middle Class in Late-Nineteenth and Early-Twentieth

Century Britain', in Marijke Gijswijt-Hofstra and Hilary Marland (eds), *Cultures of Child Health in Britain and the Netherlands in the Twentieth Century* (Amsterdam and New York: Rodopi, 2003), pp. 103–25.

Haley, Bruce, *The Healthy Body and Victorian Culture* (Cambridge, MA and London: Harvard University Press, 1978).

Hall, Lesley A., 'A Suitable Job for a Woman: Women Doctors and Birth Control to the Inception of the NHS', in Anne Hardy and Lawrence Conrad (eds), *Women and Modern Medicine* (Amsterdam and New York: Rodopi, 2001), pp. 127–47.

Hardy, Anne, *Health and Medicine in Britain since 1860* (Houndmills: Palgrave, 2001).

Hardyment, Christina, *Dream Babies: Child Care from Locke to Spock* (London: Jonathan Cape, 1983).

Hargreaves, Jennifer, *Sporting Females: Critical Issues in the History and Sociology of Women's Sports* (London: Routledge, 1994).

Harmond, Richard, 'Progress and Flight: An Interpretation of the American Cycle Craze of the 1890s', *Journal of Social History*, 5 (Winter, 1971–72), 235–57.

Harris, Bernard, *The Health of the Schoolchild: A History of the School Medical Service in England and Wales* (Buckingham and Philadelphia, PA: Open University Press, 1995).

Harrison, Barbara, 'Women and Health', in June Purvis (ed.), *Women's History: Britain, 1850–1945* (London: UCL Press, 1995), pp. 157–92.

Harrison, Barbara, *Not Only the 'Dangerous Trades': Women's Work and Health in Britain, 1880–1914* (London: Taylor & Francis, 1996).

Harrison, Brian, 'Women's Health and the Women's Movement in Britain: 1840–1940', in Charles Webster (ed.), *Biology, Medicine and Society 1840–1940* (Cambridge: Cambridge University Press, 1981), pp. 15–71.

Hendrick, Harry, *Images of Youth: Age, Class, and the Male Youth Problem, 1880–1920* (Oxford: Clarendon Press, 1990).

Hilton, Mary and Maria Nikolajeva (eds), *Contemporary Adolescent Literature and Culture* (Farnham, Surrey: Ashgate, 2012), pp. 1–16.

Hindle, Roy, *Oh, No Dear! Advice to Girls a Century Ago* (Newton Abbot: David & Charles, 1982).

Horwood, Catherine, ' "Girls Who Arouse Dangerous Passions": Women and Bathing, 1900–39', *Women's History Review*, 9 (2000), 653–73.

Hunt, Felicity (ed.), *The Schooling of Girls and Women, 1850–1950* (Oxford: Basil Blackwell, 1987).

Ineson, Antonia and Deborah Thom, 'T.N.T. Poisoning and the Employment of Women Workers in the First World War', in Paul Weindling (ed.), *The Social History of Occupational Health* (London: Croom Helm, 1985), pp. 89–107.

Jalland, Pat and John Hooper (eds), *Women from Birth to Death: The Female Life Cycle in Britain 1830–1914* (Brighton: Harvester, 1986).

Jeffs, Tony, 'Oft-Referenced – Rarely Read? Report of the Inter-Departmental Committee on Physical Deterioration 1904', in R. Gilchrist, T. Jeffs and J. Spence (eds), *Drawing on the Past: Studies in the History of Community Youth Work* (Leicester: The National Youth Agency, 2006), pp. 43–59.

Jones, Greta, *Social Hygiene in Twentieth-Century Britain* (London: Croom Helm, 1986).

Jones, Greta, 'Women and Eugenics in Britain: The Case of Mary Scharlieb, Elizabeth Sloan Chesser, and Stella Browne', *Annals of Science*, 51 (1995), 481–502.

Jones, Helen, 'Women Health Workers: The Case of the First Women Factory Inspectors in Britain', *Social History of Medicine*, 1 (1988), 165–81.

Jones, Helen, *Health and Society in Twentieth-Century Britain* (London and New York: Longman, 1994).

Jordan, Ellen, ' "Making Good Wives and Mothers"? The Transformation of Middle-Class Girls' Education in Nineteenth-Century Britain', *History of Education Quarterly*, 31 (1991), pp. 439–62.

Kamm, Josephine, *How Different from Us – Miss Buss and Miss Beale* (London: Bodley Head, 1958).

King, Helen, *The Disease of Virgins: Green Sickness, Chlorosis and the Problems of Puberty* (London and New York: Routledge, 2004).

Langhamer, Claire, *Women's Leisure in England, 1920–60* (Manchester: Manchester University Press, 2000).

Laqueur, Thomas, *Making Sex: Body and Gender from the Greeks to Freud* (Cambridge, MA and London: Harvard University Press, 1990).

Lawrence, Christopher, *Medicine in the Making of Modern Britain 1700–1920* (London and New York: Routledge, 1994).

Ledger, Sally, *The New Woman: Fiction and Feminism at the fin de siècle* (Manchester and New York: Manchester University Press, 1997).

Ledger, Sally and Roger Luckhurst (eds), *The fin de siècle: A Reader in Cultural History, c.1880–1900* (Oxford: Oxford University Press, 2000).

Lewis, Jane, *The Politics of Motherhood: Child and Maternal Welfare in England 1900–1939* (London: Croom Helm, 1980).

Lewis, Jane, *Women in England 1870–1950: Sexual Divisions and Social Change* (Brighton: Wheatsheaf, 1984).

Lewis, Jane, 'Providers, "Consumers", the State and the Delivery of Health-Care Services in Twentieth-Century Britain', in Andrew Wear (ed.), *Medicine in Society* (Cambridge: Cambridge University Press, 1992), pp. 317–46.

Livesey, Ruth, 'The Politics of Work: Feminism, Professionalization and Women Inspectors of Factories and Workshops', *Women's History Review*, 13 (2004), 233–61.

Loeb, Lori Anne, *Consuming Angels: Advertising and Victorian Women* (New York and Oxford: Oxford University Press, 1994).

Long, Vicky and Hilary Marland, 'From Danger and Motherhood to Health and Beauty: Health Advice for the Factory Girl in Early Twentieth-Century Britain', *Twentieth Century British History*, 20 (2009), pp. 454–81.

Long, Vicky, 'Industrial Homes, Domestic Factories: The Convergence of Public and Private Space in Interwar Britain', *Journal of British Studies*, 50 (2011), 434–64.

Long, Vicky, *The Rise and Fall of the Healthy Factory: The Politics of Industrial Health in Britain, 1914–60* (Houndmills: Palgrave Macmillan, 2011).

Loudon, Irvine, 'Chlorosis, Anaemia, and Anorexia Nervosa', *British Medical Journal*, 281 (20–27 December 1980), 1669–75.

Lowe, Margaret A., *Looking Good: College Women and Body Image, 1875–1930* (Baltimore, MD and London: Johns Hopkins University Press, 2003).

Malone, Carolyn, *Women's Bodies and Dangerous Trades in England, 1880–1914* (Woodbridge: Boydell, 2003).

Manton, Jo, *Elizabeth Garrett Anderson* (London: Methuen, 1965).

Marks, Patricia, *Bicycles, Bangs, and Bloomers: The New Woman in the Popular Press* (Lexington: The University Press of Kentucky, 1990).

Marland, Hilary, *Dangerous Motherhood: Insanity and Childbirth in Victorian Britain* (Houndmills: Palgrave Macmillan, 2004).

Marland, Hilary, 'Women, Health and Medicine', in Mark Jackson (ed.), *The Oxford Handbook of the History of Medicine* (Oxford: Oxford University Press, 2011), pp. 484–502.

Marland, Hilary, ' "The Diffusion of Useful Knowledge": Household Practice, Domestic Medical Guides and Medical Pluralism in Nineteenth-Century Britain', in Robert Jütte (ed.), *Medical Pluralism: Past and Present, Medizin, Gessellschaft und Geschichte Beiheft* (2013), pp. 81–100.

Martin, Jane, *Women and the Politics of Schooling in Victorian and Edwardian England* (London: Leicester University Press, 1999).

Martin, M.C., 'Gender, Religion and Recreation: Flora Lucy Freeman and Female Adolescence 1890–1925', in R. Gilchrest, T. Jeffs and J. Spence (eds), *Drawing on the Past: Studies in the History of Community Youth Work* (Leicester: The National Youth Agency, 2006), pp. 61–77.

Marwick, Arthur, *Women at War, 1914–1918* (London: Fontana, for the Imperial War Museum, 1977).

Matthews, Jill Julius, 'Building the Body Beautiful', *Australian Feminist Studies*, 5 (1987), 17–34.

Matthews, Jill, 'They had Such a Lot of Fun: The Women's League of Health and Beauty Between the Wars', *History Workshop*, 30 (1990), 22–54.

May, Jonathan, *Madame Bergman-Österberg: Pioneer of Physical Education and Games for Girls and Women* (London: George G. Harrap, 1969).

Mazumdar, Pauline M.H., *Eugenics, Human Genetics and Human Failings: The Eugenics Society, Its Sources and Its Critics in Britain* (London: Routledge, 1992).

McCrone, Kathleen E., 'Play Up! Play Up! And Play the Game! Sport at the Late Victorian Girls' Public School', *Journal of British Studies*, 23 (1984), 106–34.

McCrone, Kathleen E., *Sport and the Physical Emancipation of English Women 1870–1914* (London: Routledge, 1988).

McKibbin, Ross, *Classes and Cultures: England, 1918–1951* (Oxford: Oxford University Press, 1998).

Miller, Susan A., *Growing Girls: The Natural Origins of Girls' Organizations in America* (New Brunswick, NJ and London: Rutgers University Press, 2007).

Mitchell, Sally, *The New Girl: Girls' Culture in England 1880–1915* (New York: Columbia University Press, 1995).

Mitchinson, Wendy, *The Nature of their Bodies: Women and their Doctors in Victorian Canada* (Toronto, ON: University of Toronto Press, 1991).

Mort, Frank, *Dangerous Sexualities: Medico-Moral Politics in England since 1830* (London: Routledge & Kegan Paul, 1987).

Moscucci, Ornella, *The Science of Woman: Gynaecology and Gender in England 1800–1929* (Cambridge: Cambridge University Press, 1990).

Newman, Louise Michele, *Men's Ideas/Women's Realities: Popular Science, 1870–1915* (New York, etc.: Permagon, 1985).

Newton, Stella Mary, *Health, Art and Reason: Dress Reformers of the 19th Century* (London: John Murray, 1974).

Oppenheim, Janet, *"Shattered Nerves": Doctors, Patients, and Depression in Victorian England* (Oxford: Oxford University Press, 1991).

Pederson, Joyce Senders, 'The Reform of Women's Secondary and Higher Education: Institutional Change and Social Values in Mid and Late Victorian England', *History of Education Quarterly*, 19 (Spring 1979), 61–91.

Pickstone, John, 'Production, Community and Consumption: The Political Economy of Twentieth-Century Medicine', in Roger Cooter and John Pickstone (eds), *Companion to Medicine in the Twentieth Century* (London: Harwood Academic, 2000), pp. 1–20.

Porter, Dorothy, ' "Enemies of the Race": Biologism, Environmentalism, and Public Health in Edwardian England', *Victorian Studies*, 35 (1991), 159–78.

Porter, Dorothy, 'The Healthy Body', in Roger Cooter and John Pickstone (eds), *Companion to Medicine in the Twentieth Century* (London: Harwood Academic, 2000), pp. 201–16.

Potter, Jane, *Boys in Khaki, Girls in Print: Women's Literary Responses to the Great War 1914–1918* (Oxford: Clarendon, 2005).

Proctor, Tammy M., *Scouting for Girls: A Century of Girl Guides and Girl Scouts* (Santa Barbara, CA: Praeger, 2009).

Purvis, June, 'The Double Burden of Class and Gender in the Schooling of Working-Class Girls in Nineteenth-Century England, 1800–1870', in Len Barton and Stephen Walker (eds), *Schools, Teachers and Teaching* (Lewes: Falmer, 1981), pp. 97–116.

Purvis, June, *A History of Women's Education in England* (Milton Keynes and Philadelphia, PA: Open University Press, 1991).

Rabinbach, Anson, *The Human Motor: Energy, Fatigue and the Origins of Modernity* (Berkeley, CA: University of California Press, 1992).

Richards, Thomas, *The Commodity Culture of Victorian England: Advertising and Spectacle 1851–1914* (Stanford, CA: Stanford University Press, 1990).

Richardson, Angelique and Chris Willis (eds), *The New Woman in Fiction and in Fact: fin-de-siècle Feminisms* (Basingstoke: Palgrave, 2001).

Richardson, Angelique, *Love and Eugenics in the Late Nineteenth Century* (Oxford: Oxford University Press, 2003).

Richardson, Angelique, 'The Birth of National Hygiene and Efficiency: Women and Eugenics in Britain and America 1865–1915', in Ann Heilmann and Margaret Beetham (eds), *New Women Hybridities: Femininity, Feminism and International Consumer Culture, 1880–1930* (London and New York: Routledge, 2004), pp. 240–62.

Robbins, Ruth (ed.), *Medical Advice for Women, 1830–1915* (Abingdon: Routledge, 2008–).

Roberts, Elizabeth, *A Woman's Place: An Oral History of Working-Class Women 1890–1940* (Oxford: Basil Blackwell, 1984).

Roberts, Robert, *The Classic Slum: Salford Life in the First Quarter of the Century* (Penguin edn, 1973; first published Manchester: University of Manchester Press, 1971).

Rosenberg, Charles (ed.), *Right Living: An Anglo-American Tradition of Self-Help and Hygiene* (Baltimore, MD and London: Johns Hopkins University Press 2003).

Ross, Ellen, *Love and Toil: Motherhood in Outcast London, 1870–1918* (New York and Oxford: Oxford University Press, 1993).

Rubinstein, David, 'Cycling in the 1890s', *Victorian Studies*, 21 (Autumn, 1977), 47–71.

Russett, Cynthia Eagle, *Sexual Science: The Victorian Construction of Womanhood* (Cambridge, MA: Harvard University Press, 1989).

Schiebinger, Londa, *The Mind has No Sex? Women in the Origins of Modern Science* (Cambridge, MA and London: Harvard University Press, 1989).

Schwartz, Hillel, *Never Satisfied: A Cultural History of Diets, Fantasies and Fat* (New York: Free Press, 1986).

Scrimgeour, R.M. (ed.), *The North London Collegiate School 1850–1950: A Hundred Years of Girls' Education* (London, New York and Sydney: Oxford University Press, 1950).

Searle, G.R., *The Quest for National Efficiency: A Study in British Politics and Political Thought, 1899–1914* (Oxford: Blackwell, 1971).

Searle, G.R., *Eugenics and Politics in Britain, 1900–14* (Leyden: Noordhoff, 1976).

Showalter, Elaine and English Showalter, 'Victorian Women and Menstruation', in Martha Vicinus (ed.), *Suffer and be Still: Women in the Victorian Age* (London: Methuen, 1980; first published Bloomington, IN: Indiana University Press, 1972), pp. 38–44.

Showalter, Elaine, *The Female Malady: Women, Madness and English Culture, 1830–1980* (London: Virago, 1987, first published New York: Pantheon, 1985).

Skelding, Hilary, 'Every Girl's Best Friend? The *Girl's Own Paper* and its Readers', in Emma Liggins and Daniel Duffy (eds), *Feminist Readings of Victorian Popular Texts: Divergent Femininities* (Aldershot: Ashgate, 2001), pp. 35–52.

Smith, Michelle J., *Empire in British Girls' Literature and Culture: Imperial Girls, 1880–1915* (Houndmills: Palgrave Macmillan, 2011).

Smith-Rosenberg, Carroll, 'The Hysterical Woman: Sex Roles and Role Conflict in 19th-Century America', *Social Research*, 39 (1979), 652–78.

Søland, Birgitte, *Becoming Modern: Young Women and the Reconstruction of Womanhood in the 1920s* (Princeton, NJ and Oxford: Princeton University Press, 2000).

Soloway, R.A., *Demography and Degeneration: Eugenics and the Declining Birthrate in Twentieth Century Britain* (Chapel Hill, NC: University of North Carolina Press, 1990).

Springhall, John, *Youth, Empire and Society: British Youth Movements 1883–1940* (London: Croom Helm, 1977).

Steadman, F. Cecily, *In the Days of Miss Beale: A Study of Her Work and Influence* (London: E. J. Burrow, 1936).

Stearns, Carol Zisowitz and Peter N. Stearns, *Anger: The Struggle for Emotional Control in America's History* (Chicago and London: University of Chicago Press, 1986).

Steedman, Carolyn, *Childhood, Culture and Class in Britain: Margaret McMillan, 1860–1931* (London: Virago, 1990).

Steedman, Carolyn, 'Bodies, Figures and Physiology: Margaret McMillan and the Late Nineteenth-Century Remaking of Working-Class Childhood', in Roger Cooter (ed.), *In the Name of the Child: Health and Welfare 1880–1940* (London and New York: Routledge, 1992), pp. 19–44.

Steel, Valerie, *The Corset: A Cultural History* (New Haven, CT and London: Yale University Press, 2001).

Stephen, Barbara, *Emily Davies and Girton College* (London: Constable, 1927).

Stewart, Mary Lynn, *For Health and Beauty: Physical Culture for Frenchwomen, 1880s–1930s* (Baltimore, MD and London: Johns Hopkins University Press, 2001).

Strange, Julie-Marie, 'The Assault on Ignorance: Teaching Menstrual Etiquette in England, c.1920s to 1960s', *Social History of Medicine*, 14 (2001), 247–65.

Summers, Leigh, *Bound to Please: A History of the Victorian Corset* (Oxford: Berg, 2001).

Taylor, Harvey, *A Claim on the Countryside: A History of the British Outdoor Movement* (Edinburgh: Keele University Press, 1997).

Theriot, Nancy M., 'Negotiating Illness: Doctors, Patients, and Families in the Nineteenth Century', *Journal of the History of the Behavioral Sciences*, 37 (2001), 349–68.

Thom, Deborah, *Nice Girls and Rude Girls: Women Workers in World War I* (London: I.B. Taurus, 1997).

Thomson, Mathew, *Psychological Subjects: Identity, Culture, and Health in Twentieth-Century Britain* (Oxford: Oxford University Press, 2006).

Tinkler, Penny, *Constructing Girlhood: Popular Magazines for Girls Growing Up in England 1920–1950* (London: Taylor & Francis, 1995).

Todd, Selina, *Young Women, Work and Family 1918–50* (Oxford: Oxford University Press, 2005).

Tomes, Nancy, *The Gospel of Germs: Men, Women and the Microbe in American Life* (Cambridge, MA and London: Harvard University Press, 1998).

Ueyama, Takahiro, *Health in the Marketplace: Professionalism, Therapeutic Desires, and Medical Commodification in Late-Victorian London* (Palo Alto, CA: Society for the Promotion of Science and Scholarship, 2010).

Valenze, Deborah, *The First Industrial Woman* (New York and Oxford: Oxford University Press, 1995).

Verbrugge, Martha H., *Active Bodies: A History of Women's Physical Education in Twentieth-Century America* (Oxford and New York: Oxford University Press, 2012).

Vernon, James, *Hunger: A Modern History* (Cambridge, MA and London: Harvard University Press, 2007).

Vertinsky, Patricia E., 'Body Shapes: The Role of the Medical Establishment in Informing Female Exercise and Physical Education in Nineteenth-century North America', in J.A. Mangan and Roberta J. Parke (eds), *From 'Fair Sex' to Feminism: Sport and the Socialization of Women in the Industrial and Post-Industrial Eras* (London: Frank Cass, 1987), pp. 256–81.

Vertinsky, Patricia E., *The Eternally Wounded Woman: Women, Doctors, and Exercise in the Late Nineteenth Century* (Urbana and Chicago, IL: University of Illinois Press, 1994; first published Manchester: Manchester University Press, 1989).

Vester, Katharina, 'Regime Change: Gender, Class, and the Invention of Dieting in Post-Bellum America', *Journal of Social History*, 44 (2010), 39–70.

Welshman, John, 'Child Health, National Fitness, and Physical Education in Britain, 1900–1940', in Marijke Gijswijt-Hofstra and Hilary Marland (eds.), *Cultures of Child Health in Britain and the Netherlands in the Twentieth Century* (Amsterdam and New York: Rodopi, 2003), pp. 61–84.

Whorton, James C., 'The Hygiene of the Wheel: An Episode in Victorian Sanitary Science', *Bulletin of the History of Medicine*, 52 (1978), 61–88.

Whorton, James C., *Crusaders for Fitness: The History of American Health Reformers* (Princeton, NJ: Princeton University Press, 1982).

Whorton, James C., *Inner Hygiene: Constipation and the Pursuit of Health in Modern Society* (Oxford: Oxford University Press, 2000).

Winter, J.M., *The Great War and the British People* (Houndmills: Macmillan, 1985).

Wood, Jane, *Passion and Pathology in Victorian Fiction* (Oxford: Oxford University Press, 2001).

Woollacott, Angela, ' "Khaki Fever" and its Control: Gender, Class, Age and Sexual Morality on the British Homefront in the First World War', *Journal of Contemporary History*, 29 (1994), 325–47.

Woollacott, Angela, 'Maternalism, Professionalism and Industrial Welfare Supervisors in World War One Britain', *Women's History Review*, 3 (1994), 29–56.

Woollacott, Angela, *On Her Their Lives Depend. Munitions Workers in the Great War* (Berkeley, CA and London: University of California Press, 1994).

Zelizer, V.A., *Pricing the Priceless Child: The Changing Social Value of Children* (New York: Basic Books, 1985).

Zweiniger-Bargielowska, Ina, 'Raising a Nation of "Good Animals": The New Health Society and Health Education Campaigns in Interwar Britain', *Social History of Medicine*, 20 (2007), pp. 73–89.

Zweiniger-Bargielowska, Ina, *Managing the Body: Beauty, Health, and Fitness in Britain, 1880–1939* (Oxford: Oxford University Press, 2010).

Unpublished theses

Aspinall, Susan, 'Nurture as well as Nature: Environmentalism in Representations of Women and Exercise in Britain from the 1880s to the Early 1920s', unpublished University of Warwick PhD thesis, 2008.

Collins, Tracy J.R., 'Physical Fitness, Sports, Athletics and the Rise of the New Woman', unpublished University of Purdue PhD thesis, 2007.

Dove, Iris, 'Sisterhood or Surveillance? The Development of Working Girls' Clubs in London 1880–1939', unpublished University of Greenwich PhD thesis, 1996.

Rutter, Jan, 'The Young Women's Christian Association of Great Britain, 1900–1925: An Organisation of Change', unpublished University of Warwick, MA Dissertation', 1986.

Stanley, Gregory Kent, 'Redefining Health: The Rise and Fall of the Sportswoman. A Survey of Health and Fitness Advice for Women, 1860–1940', unpublished University of Kentucky, PhD thesis, 1991.

Electronic resources

British Periodicals Collection

Chadwyck Gerritsen Collection

Defining Gender, 1450–1910, Matthew Adams Digital, Bodleian Library

Dictionary of National Biography

Index

adolescence, 6, 17, 35–9, 40–1, 151–2, 169
advice literature, 1–2, 9, 42–85, 91–2, 96, 97–102, 159–60, 190–4
  cycling, 91–2, 103–4
  direct to girls, 53–4
  feminist movement, 43, 90
  hydropathy, 53
  impact of, 84–5
  physical culture, 71, 82, 100, 119
  physical fitness, 45–6, 51–3
Alexander, A., 129, 167
anaemia, 26, 54–5, 58–9, 65
  in working-class girls, 156, 174, 183–4
Anderson, Elizabeth Garrett (Dr), 10, 12, 20, 133, 139, 142, 151
Andrews, Bim, 96, 115–16, 190
anorexia nervosa, 6, 16, 30–3, 72
Arnold, Wallace, 81, 82
Association of Assistant Mistresses, 145–6
Association of Teachers of Domestic Science, 148
Atkinson, Paul, 124, 142
attitude, positive, 62–5
'auto-intoxication', 72

Ballin, Ada, 97, 115
Barnard, Amy, 64, 67, 85, 98–9, 102, 120
Baron, Barclay, 169–70
Beale, Dorothea, 134, 142, 144–5, 147
beauty, 43, 49, 65–8
Beier, Lucinda McCray, 159, 165
Bergman Österberg, Martina, 88, 95, 125
'bicycle face girl', 11, 86, 105
'biological determinism', 17

biological vulnerability, 6–7, 10, 12, 15–41, 131–42
birth control, 8
birth rate, falling, 35, 120
Black, George (Dr), 23, 55–6
Blackwell, Elizabeth (Dr), 134–5
Blanchard, Phyllis, 36
Bourne, E.D., 93
boy's health, 21–2, 29, 93, 141
  see also male adolescence
Braybon, Gail, 178
breathing exercises, 60–1, 63, 67–8, 90, 98, 167–8
British Gynaecological Society, 34
British Medical Journal (BMJ), 88, 105, 136, 138
Broome, Mary Anne, 3
Brumberg, Joan Jacobs, 31
Bryant, Sophie, 142–3
Buckley, Cheryl, 71
Burstall, Sara, 147–9
Buss, Francis Mary, 128, 130, 133, 134, 135, 142–3, 145, 234 n 90

Caulfeild, S.F.A., 168
Chant, Laura Ormiston, 87
Chavasse, Pye Henry (Dr), 54, 58, 63, 91, 215 n 44
Cheltenham Ladies College, 134, 144–5, 151
Chesser, Elizabeth Sloan (Dr), 38–9, 53, 61, 62, 64, 69, 75, 84, 148, 150–2, 161, 174, 193–4, 195
Childhood Society, 36
Chisholm, Catherine (Dr), 124, 149, 151
chlorosis, 6, 25–6, 54, 58–9, 65
Chreiman, M.A. (Miss), and benefits of exercise, 127–9

citizenship, 35, 40–1, 181–2, 186–7, 188, 192–3
  see also national efficiency
Clark, Andrew (Dr), 25
Clarke, Edward H. (Dr), 18–20
clitoridectomy, 34
clothing, 70, 73–5, 114–5, 116, 147
  for cycling 114–16
Clouston, Thomas (Dr), 12, 21, 135–6
clubs, girls', 8–9, 161–77, 190
co-education, 19
Collier, D. J., 183
Corke, Helen, 57
corsets, 33–4, 39, 50, 73–5, 110, 146, 166
  and damage to health, 33–4, 73–5, 166
  in *Girl's Own Paper*, 78–9
Countess of Malmesbury, 107, 117
Crichton-Browne, James (Dr), 12, 139–40
Curwen, Maud, 45, 58–9, 60, 65, 67, 160
cycling, 46, 79–80, 90–1, 103–18
  advice literature, 91–2, 103–4
  critics, 86–8, 104, 120–1
  dress for, 114–16
  health benefits of, 79–80, 87, 89, 108–14
  health concerns, 105–7
  and reproduction, 87, 88, 105–7
  and sexual stimulation, 105–6, 226 n 94
  as transformative, 117–18

Dangerous Trades legislation, 157
Davidson, Lillias Campbell, 116
Davin, Anna, 156
dental hygiene, 166
Deptford Camp School, 126–7
de-sexing, 78, 87–8, 191
Dickinson, Robert (Dr), 106
diet, 67, 71–3, 165–6, 174
domestic 'science', 72, 147–8, 158, 165
  see also housework, as exercise
Donkin, H. B. (Dr), 28–9
'double burden', 156, 169, 196
'double gain', 191, 196

Dove, Jane Frances, 130, 146–7
Duffin, Lorna, 6
Dukes, Clement (Dr), 62, 140–1, 152
  health and schooling, 129–30
Dyhouse, Carol, 8, 17, 36, 44, 47

Education Acts, 131, 133
education, 3–4, 5
  higher, 10–11, 17–21, 24, 29–30, 123–4, 128, 131–42, 146–8
  Empire concerns, 119, 152, 188, 192–3
  see also citizenship
energy, fixed funds of, 17–21, 61, 81, 92, 123, 148–9, 189
Ethel Herbert, 45, 65, 67
Eugenics Society, 38
eugenics, 35, 38, 151–2, 169
  'feminist eugenicists', 53, 61, 83, 87–8, 120, 149–51, 191
Everett-Green, Constance, 104, 105
exercise and sport, 43, 69–71, 81–2, 86–121, 124, 125, 127–9, 130–31, 142–3, 145–7, 149, 158–9, 162–3, 167–71
  and character, 102–3, 145, 150, 170
  see also physical fitness

factories
  factory inspectors, 18, 53, 158
  legislation, 13
  'factory girl', 157, 169, 173–4
  see also working-class girls
Farningham, Marianne, 59, 119
Fawcett, Hilary, 71
feminist movement, 20, 38, 89–90, 134–5
  magazines, 43, 90
Fenton, W. H. (Dr), 109, 110, 112, 114
Ferris, Helen J., 187
Findlay, M. E., 36–7, 151–2
First World War, 3, 7–8, 35, 38, 40–1, 159, 178–85
'fixed fund of energy' theory, 17–21, 61, 81, 92, 123, 148–9, 189
'flapper', 2, 71, 184
Fletcher, J. Hamilton, 92–3, 102
Fletcher, Sheila, 90
Flint, Kate, 19, 139
Fortnightly Review, 12, 19, 139

Foucault, Michel, 41
Frawley, Maria H., 6
Freeman, Flora Lucy, 170, 177
Froebel Society, 125

Galabin, A.L. (Dr), 26
Galbraith, Anna M. (Dr), 52
giant stride, 131, 132
Girl Guides, 169–70, 171, 177, 184, 190
'Girl of the Period', 11, 45–7
Girl's Own Paper (GOP), 2, 4, 42–3, 48, 75–82, 159
    corsets, 78–9
    Gordon Stables in, 2, 4, 79–81, 82, 110, 159
    hysteria, 79
Girl's Realm, 48, 90, 100, 168
'girlhood', 3, 4, 49–51, 59
    for middle-class girls, 4–5
    for working-class girls, 4
Girls' Brigade, 168
Girls' Friendly Society (GFS), 168, 170, 177
'Girton Girl', 10–11, 48, 123
Gorham, Deborah, 60
Grand, Sarah, 33
Greta, Jones, 195
Gull, William (Dr), 30–2, 137
gymnastics, 69–71, 80, 90, 95, 96, 98, 100–01, 128–9, 130–1, 143, 146, 153, 167
    as remedial, 126, 129, 130–1
gynaecology, 17

Haley, Bruce, 88
Hall, G. Stanley, 6, 35–7, 38
Hallam, Margaret, 59, 65, 68
Harrison, Barbara, 7, 53, 157
Harrison, Brian, 53
Headmistresses Association, 145–6
health advice, 1–2, 12–13, 38, 41, 42–88, 159–60
health periodicals, 51
Herbert, Ethel, 45, 58–9, 60, 67, 160
'heredity', 6
Herrick, Christine Terhune, 102–3

Hoggan, Frances (Dr), 106–7, 122, 124, 127–8, 130, 134
    and benefits of exercise, 127–8
    on cycling, 106–7
Hooper, John, 33
Horwood, Catherine, 90
housework, as exercise, 68–9, 96, 98, 176
    see also domestic 'science'
Humphreys, Mary, 53, 63, 68, 75, 76, 102, 193, 194
hydropathy, 27
    advice literature on, 53
hygiene, 8, 67–9, 114–15, 167, 183
    sexual, 68, 83–4, 184–5
    of working girls, 166, 183
hysteria, 6, 19, 27–30, 64–5
    in *Girl's Own Paper*, 79

industrial diseases, 156, 178
Industrial Fatigue Board, 186
infants, care of, 69, 120, 148, 176, 182, 192
intemperance, 84, 164
Interdepartmental Committee on Physical Deterioration (1904), 169

Jalland, Pat, 33
Johnson, Edward (Dr), 26–7
Johnson, Walter (Dr), 27–8
Jones, Robert (Dr), 153
journals, *see* advice literature

Kellerman, Annette, 83
Kelly, Howard A. (Dr), 33–4, 39, 137, 144–5
Kelynack, T.N. (Dr), 37
Kenealy, Arabella (Dr), 37, 53, 61, 86, 97
    critics of, 87, 120
kindergarten training, 125
Kingsford, Anna Bonus (Dr), 65–7, 72
Kinnaird, Mary, 162
Kirk, Edward Bruce, 57–8
Kitching, Mavis, 171

Lancet, 31, 34, 88, 150
Langhamer, Claire, 4, 173
'Lawn Tennis Girl', 40, 86

Lesège, Charles (Dr), 30
Ling system of 'free exercises', 125
Linton, Eliza Lynn, 11, 46, 104
Liston, Lawrence (Dr), 82, 99, 111
little mothers, 9
Liverpool Gymnasium, 129, 167
London School Board, 125, 127
Long, Vicky, 158, 160, 180
Loudon, Irvine, 25, 156
Lowe, Ernest W., 94–6

MacDonald, Sarah, 183–4
MacFadden, Bernarr, 69
MacNaughton-Jones, H. (Dr), 34
male adolescence, 21–2, 34, 35, 145
   see also boy's health
Malone, Carolyn, 157
Manchester High School for Girls,
   124, 147
'Manly Maidens', 88
Martineau, Harriet, 89–90
masturbation, 34
maternity, 7–8, 49, 145, 157, 193
   see also motherhood
Matthews, Jill Julius, 117
Maudsley, Henry (Dr), 10, 12, 18–20,
   24, 29, 39, 132–3, 135, 139, 148
McCrone, Kathleen E., 90, 117, 147
McMillan, Margaret, 126
medical inspections, 130–1, 144–5,
   149, 158–9
Medical Women's Federation, 6, 150,
   151, 186, 195
'Medicus', *see* Stables, Gordon (Dr)
menstruation, 5, 6, 16, 22, 25, 36
   and curtailment of physical
      exercise, 55, 151
   disorders of, 25–7, 36–7, 55–6,
      136–7, 183
   hygiene, 26, 54, 59
   retardation of, 22–3, 136
   role of mothers in managing, 54,
      57, 136
   see also puberty
Michelet, Jules, 16
middle-class girls, 4–5, 18, 123–4, 133,
   186
   hardships faced by, 9, 123

Mitchell, Sally, 3–4, 10, 30, 98, 119,
   190
Mitchell, Silas Weir (Dr), 36
Moore, William Withers (Dr), 20–1
Morris, Malcolm (Dr), 74, 140, 170
Mortimer-Granville, J. (Dr), 21–2
   and electric vibrator, 204–5 n 26
motherhood, 10, 38, 182, 191–5
   and mental fitness, 136, 141–2, 150,
      151–2
   and physical fitness, 11, 24, 94–6,
      119–20, 123, 150
   see also maternity
Mowbray House Cycling Association,
   111, 112–13
Munition Workers' Welfare
   Committee, 179
Murrell, Christine M. (Dr), 151, 195

National Cyclists Union, 105
national efficiency, 38
   see also citizenship
National Federation of Women
   Workers, 162
National Health Society, 134–5
National Organisation of Girls Clubs
   (NOGC), 172
National Union of Women Workers,
   162
nerve energy, 18–21, 61
nerves, management of, 60, 62–5
neuralgia, 19
neurasthenia, 30
'New Girl', 3, 46
New Journalism, 47
'New Woman', 11, 12, 30, 40, 46, 88,
   89, 90, 116
Newsholme, Arthur (Dr), 8
North London Collegiate School for
   Girls (NLCS), 122, 128–9, 143–4,
   153–4
   medical inspections, 130–1
Northcliffe (Lord), 49

ovariotomy, 20
overexertion, 61, 88, 89–103, 109–11,
   113, 140–2, 148–52, 191
overstudy, 19–21, 24, 30, 37, 135–42

Parkes, Bessie Rayner, 90
Peters, Charles, 98, 159
Pfeiffer, Emily, 137
physical culture, 69–75, 100, 101, 167
  advice literature, 72, 100
  see also physical fitness
physical fitness, 20, 26–7, 40
  concerns about lack of, 92–5
  housework and, 68, 96
  literature advice, 45–6, 51–3
  moderation of, 96–7, 148–9
  promotion of, 89–103
  'risks' to femininity, 86–8, 92,
    120–1
  working girls, 96
  see also exercise and physical
    culture
'physical religion', 127
Pickstone, John, 7–8
Playfair, William Smout (Dr), 28, 31,
  40, 86, 138–9, 145
Porter, Dorothy, 38
posture, 63
poverty, 8–9, 125–7, 156, 157–8
  see also working-class girls
Pritchard, Eric, Mrs, 120, 191
psychology, feminisation of, 62
puberty, 5–6, 53–9
  'perils of', 21–35
  see also menstruation
Purvis, June, 156, 169

Queen, 51, 103–4

Rabinbach, Anson, 114
Rentoul, Robert Reid (Dr), 23–4
reproductive systems, 7–8
  women as prisoners of, 17
Richards, Florence Harvey (Dr), 52
Richards, Thomas, 49
Richardson, Benjamin Ward (Sir),
  97–8, 109–11
Robarts, Emma, 162
Roberts, Elizabeth, 4
Roberts, Robert, 4, 157
Rock, Shadwell, 33
Ross, Ellen, 8–9
Rubinstein, David, 103
Ruddock, Edward H. (Dr), 27

Sandow, Eugen, 69, 89
Scharlieb, Mary (Dr), 6, 38–9, 53,
  83–4, 149–50, 152, 184–5, 191,
  193, 195
Schofield, Alfred T. (Dr), 64, 93–4, 106,
  114
school medical service, 127, 158
sexual hygiene, 68, 83–4, 184–5
Shadwell, A. (Dr), 88, 106
Showalter, Elaine, 33
Skey, F.C. (Dr), 28
Slaughter, J.W., 37
sleep, 65
slimming craze, 71
Smith, Annie Burns, 69, 74–5, 192
Smith, Michelle J., 148, 190
South African (Anglo-Boer) War, 35,
  120, 169, 191
Spencer, Herbert, 18, 203 n 13
spinal curvature, 122, 129, 130, 131,
  141, 145, 149, 169
Stables, Gordon (Dr), 1–2, 4, 12, 51–2,
  56–7, 60–1, 62, 73, 79–82, 159, 160
  on beauty, 66–7
  changing views on sport, 120
  Girl's Own Paper (GOP), 2, 4, 79–81,
    82, 110, 159
  publications of, 1, 51–2
Stanley, Maude, 172, 174, 176
Stearns, Peter N. and Carol, 44
Steedman, Carolyn, 195
Stenhouse, Agnes L. and E. 61, 67, 192
Stewart, Mary Lynn, 16
Strange, Julie-Marie, 19
suffragist, 12
Swedish system of 'free exercises', 125

Tait, Robert Lawson (Dr), 34
Thackrah, Margaret (Dr), G., 150
Theriot, Nancy, 40
Thomson, Mathew, 62
Thorburn, John (Dr), 12, 24, 26, 33,
  136–7
tight-lacing, *see* corsets
Tilt, Edward (Dr), 12, 22–3, 54, 136
Tinkler, Penny, 77, 159
Treves, Frederick (Dr), 33–4
Turner, Edward Beardon (Dr), 105,
  107, 108, 112

uterine disease, 19

Valenze, Deborah, 173
venereal diseases, 83–4, 178–9, 184–5
Vertinsky, Patricia E., 6, 16
Vester, Katharina, 72

Walker, Emma (Dr), 59, 60, 91–2, 97
Walsh, J.H. (Dr), 54
Weatherly, Lionel (Dr), 54, 55, 83
Webb, Beatrice, 170, 182–3
'white blouse' revolution, 3
Whitley, Mary, 45–6, 97
Whorton, James C., 88, 105
Wollstonecraft, Mary, 11
women doctors, as authors of
    literature advice, 53
Women's Co-operative Guild, 8
working-class girls, 157–8, 161–77
    advice literature, 60, 159–60
    behaviour, 160, 180, 182
    cycling, 105, 112–3
    diet, 164–6, 176, 179–81
    employment of, 3, 161, 178
    exercise and sport, 96, 158–9, 170–1
    First World War, 178–85
    'girlhood', 4
    hardships faced by, 9, 157–8, 160
    health, 156–61
    housing, 157–8

hygiene, 167, 183
impact of poverty, 18
and monotony, 162, 174, 183, 184,
    186–7
and YWCA, *see* Young Women's
    Christian Association (YWCA)
    see also 'factory girl'
Wycombe Abbey School, 146

Young Women's Christian Association
    (YWCA), 9, 13, 51, 155–6, 161–77
    advice literature, 162–3, 165–6
    canteens, 166, 179–81
    convalescence and holidays,
        171–2
    and diet, 164–6, 176, 179–81
    exercise and sport, 162–3, 167–71
    health of girls, 162–7, 171
    Industrial Law Committee, 175
    Moral Education Sub-Committee,
        184
    and moral and religious wellbeing
        of girls, 155, 162, 164, 166, 167,
        177, 184
youth, *see* adolescence
YWCA, *see* Young Women's Christian
    Association (YWCA)

Zelizer, Viviana, 195
Zweiniger-Bargielowska, Ina, 48, 71

Printed in the United States
By Bookmasters